CASS LIBRARY OF AFRICAN STUDIES

GENERAL STUDIES

No. 72

Editorial Adviser: JOHN RALPH WILLIS

Centre of West African Studies, University of Birmingham

MUSHAMALENGI, "A ROYAL PRINCE" OF THE BAKUBA KINGDOM
IN THE UPPER KASAI.

DAWN
IN DARKEST AFRICA

BY

JOHN H. HARRIS

With an Introduction
by
THE RIGHT HON. THE EARL OF CROMER

FRANK CASS & CO. LTD.
1968

Published by
FRANK CASS AND COMPANY LIMITED
67 Great Russell Street, London WC1
by arrangement with John Murray (Publishers) Ltd.

First edition   1912
New impression   1968

SBN 7146 1672 9

Printed in Great Britain by
Thomas Nelson (Printers) Ltd., London and Edinburgh

INTRODUCTION

By the Earl of Cromer

I HAVE been asked to write a short introduction to this book, and I have no hesitation in complying with the request.

Although the high motives and disinterested devotion which inspire missionary and philanthropic effort are very generally recognized, there is often a predisposition —more frequently felt than expressed—not only amongst responsible officials but also in the minds of no inconsiderable portion of the public to accept with some reserve both the accuracy of the facts and the soundness of the conclusions emanating wholly from these sources. This scepticism, provided it be not allowed to degenerate into unworthy prejudice, is not merely healthy but even commendable. I could mention cases within my own knowledge where missionary zeal was certainly allowed to outrun discretion. It is the duty of responsible officials to be sceptical in such matters. Whilst sympathizing with humanitarians they should endeavour to remedy whatever of quixotism is to be found in their suggestions; and to guide those from whom those suggestions emanate along a path calculated to ensure the achievement of their objects by the adoption of practical methods which will be consonant with the moral and material interests of the Empire at large.

Occasional errors, the result of unchecked enthusiasm in a noble cause, cannot, however, for one moment be allowed to outweigh the immense benefits conferred on civilization by missionary and philanthropic agencies. Nowhere have these benefits been more conspicuous than in the case of the Congo.

The fact that but a few years ago the administration of the Congo was a disgrace to civilized Europe is now so fully recognized, not only in this country, but also— to the honour of the Belgians be it said—in Belgium itself, that it is scarcely necessary to labour the point. One startling fact is sufficient to demonstrate its true character. According to an estimate made by Sir Reginald Wingate,* the population of the Soudan under the Mahdi's rule was reduced from 8,525,000 to 1,870,500 persons ; in other words over 75 per cent. of the inhabitants died from disease or were killed in external or internal wars. The civilized European who for some years presided over the destinies of the Congo was no more merciful, save as a matter of percentage, than the ignorant and fanatical Dervish at Khartoum. Mr. Harris states (p. 208) that, under the regime of King Leopold, the Congo population was reduced from 20,000,000 to 8,000,000.† More than this. It is generally impossible in the long run to pronounce a complete divorce between moral and material interests. It will, therefore, be no matter for surprise that the Leopoldian policy was as unsuccessful from an economic as it was from an humanitarian point of view. It is now clear that unbridled

* *Parl. Paper Egypt*, No. 1 of 1904, p. 79.
† Stanley estimated the whole of the Congo population at 40,000,000.

company-mongering has gone far to destroy the sources of wealth to which it owed its birth. Mr. Harris tells a piteous tale of the manner in which the rubber vines have been handled, and, generally of the condition of the plantations. Neither, having regard to the wanton destruction of elephants (p. 213) does ivory appear to have been much more tenderly treated than rubber.

Even the most hardened sceptic as regards the utility of missionary enterprise will not, I think, be prepared to deny that to the Missionaries, in conjunction with Mr. Morel, the main credit accrues of having brought home to the British public, and eventually to the public of Europe, the iniquities which, but a short time ago, were being practised under European sanction in the heart of Africa.

Amongst this devoted band, many of whom have paid with their lives the heavy toll which cruel Africa exacts, none have been more steadfast in their determination to insist on the reform of the Congo administration than the writer of this book—Mr. Harris. None, moreover, have brought a more evenly-balanced mind to bear on the numerous problems which perplex the African administrator. Mr. Harris may be an enthusiast, but of this I am well convinced—both by frequent personal intercourse and from a careful perusal of his work— that his enthusiasm is tempered by reason and by a solid appreciation of the difference between the ideal and the practical. He wisely (p. 35) deprecates undue Missionary interference with local customs. He has even something (pp. 58–60) to say in palliation of polygamy, and if I rightly understand his remarks on p. 154, he does not utterly exclude a resort to forced labour under certain conditions and under certain circumstances.

Moreover, in so far as my experience enables me to form an opinion, Mr. Harris has acquired a firm grasp of the main principles which should guide Europeans who are called upon to rule over a backward and primitive society, and of the fact that prolonged neglect of those principles must sooner or later lead to failure or even disaster. He writes as a fair-minded and thoroughly well-informed observer. Throughout his pages may be found many acute observations on the various problems which, in forms more or less identical, tax the ingenuity of the governing race wherever the white and the coloured man meet as ruler and subject. Notably Mr. Harris dwells (p. 67) on the great influence exerted by the example set by officials ; this example, he thinks—most rightly in my opinion—is more important than the issue of laws and decrees. Here, he says—and I quote the passage with regret—" is where the Belgian and French Congo officials have failed so utterly." To put the matter in another, and somewhat mathematical, form, I have always held that 75 per cent. of the influence of British officials for good depends on character, and only 25 per cent. on brains. Mistakes arising from defective intelligence will generally admit of being rectified. Those which are due to defects of character are more often irremediable. My belief is that the great and well-deserved success which has attended Sir Reginald Wingate's administration of the Soudan arises in no small degree from a recognition of this commonplace, and from its practical application in the choice of officials.

I am not sure that its importance is always adequately recognized in London. It is well to encourage the importation of cocoa and palm oil into the London market. But it is better to acquire the reputation, which (p. 280) Mr. Harris says has passed into a proverb in the Congo, that " the Englishman never lies."

For these reasons I have no hesitation in recommending this book to the public. Mr. Harris' facts may perhaps be called in question by others possessing greater local knowledge than any to which I can pretend. His conclusions—notably those in his final chapter in which he re-arranges the map of Africa in a somewhat daring spirit—manifestly admit of wide differences of opinion. But he speaks with an unique knowledge of his subject. The opportunities which, with praiseworthy zeal, he and his devoted companion made for themselves to acquire a real knowledge of African affairs have been exceptional. He has thus produced a book in which the ordinary reader cannot fail to be interested if it be only by reason of the vivid and picturesque account it gives of African life and travel, and in which those who have paid special attention to African administration will find many useful indications of the directions in which their efforts towards reform may best be applied. Whatever may be thought of some of Mr. Harris' suggestions, it cannot but be an advantage, more especially now that attention is being more and more drawn to African affairs, that the Government, Parliament, and the general public should learn what one so eminently qualified as Mr. Harris to instruct them in the facts of the case has to say on the subject.

Mr. Harris is not sparing in his criticisms, neither does he withhold praise when he considers it is due. Whilst strongly condemning the slavery—for such it virtually is—that the Government of Portugal permits in its Colonies, he dwells (p. 296) on the " kindly nature " of the Portuguese themselves, and significantly adds " there is no colour-bar in the Portuguese dominions." * He appears to find little to commend in French administration, and much (pp. 90 and 91) to condemn in their commercial policy. He does justice (p. 88) to the thoroughness and wisdom of the Germans in all matters connected with trade, and does not, as I venture to think, detract from the merits of the liberal policy which they have adopted by alluding to the fact that it is based on self-interested motives. On the other hand, he strongly condemns (p. 142) the German treatment of the natives. He dwells (p. 92) on the petty and vexatious obstacles placed in the way of trade by the Belgian officials of the Congo, of which " even Belgian merchants complain," and he has, of course, little to say in favour of Belgian administration in other respects. But he has the fairness to admit (p. 209) that since the annexation of the Congo by Belgium the death rate has diminished and the birth rate increased—a fact which, after the experiences of the Leopoldian regime, appears to me to be very eloquent, and to reflect much credit on the

* There can, I can conceive, be little doubt that colour prejudice is a much greater obstacle to social intercourse in the case of the Teutonic—certainly in that of the Anglo-Saxon—races than it is in the case of Latin Europeans. Mr. Bryce (*American Commonwealth*, Vol. II., p. 355) says : " In Latin America, whoever is not black is white ; in Teutonic America, whoever is not white is black."

Belgian Government. Moreover, he tells us that " where-ever the Belgian reforms have been most completely applied, there the ravages of sleeping sickness appear to be more or less checked."

These observations are interesting, as they enable a comparison to be made with the results obtained under different systems of government, but they deal with matters for which—save to a limited degree in the Congo, and also perhaps to some slight extent as regards the continuance of slavery in the Portuguese possessions —neither the British Government nor the British nation are in any degree responsible. The internal policy to be adopted in the African territories possessed by France and Germany is a matter solely for the consideration of Frenchmen and Germans. But Mr. Harris has a good deal to say about the conduct of affairs in British African possessions, and it will be well if public attention is directed to his remarks in this connection, lest having preached to others we may ourselves become castaways.

Mr. Harris' position is so completely detached that he may, without the least hesitation, be acquitted of any desire to exalt unduly the achievements of his own countrymen. The spirit in which he writes is not national, but cosmopolitan. Moreover, he is manifestly not greatly enamoured of the proceedings of some, at all events, of the British officials. For these reasons his testimony is all the more valuable when he speaks, as is frequently the case, in terms of warm praise of the successful results which have been attained under British administration. He says it is not only the best in West Central Africa, but that the natives themselves recognize that it is the best. This testimony is all the more satisfactory because the excellence of British rule has not been always fully recognized in those circles in which Mr. Harris principally moves. " With inherent instinct," he says (p. 257), " the British Government recognizes that the real asset of the Colony (*i.e.* the Gold Coast Colony) is the indigenous inhabitant, whose material and moral progress is not only the first, but the truest interest of the State." It is by proceeding on this sound principle that the natives have been kept in possession of the land. " The almost phenomenal success of the cocoa industry in the British Colony of the Gold Coast," Mr. Harris says (p. 161), " is due entirely to the fact that the natives are the proprietors of the cocoa farms." It is also by the adoption of this principle that it has been possible to solve the thorny labour question. " The native farmers of Southern Nigeria and the Gold Coast employ a good deal of native labour and generally speaking find little difficulty in obtaining all they want " (p. 262).* Mr. Harris claims (p. 264) that the economic will be no less satisfactory than the moral results of the liberal policy which has been adopted by Great Britain, and that " the indigenous industry of the British Colonies working in its own interests, unencumbered by the heavy cost of European supervision and the drawbacks of imported contract labour, will, under the guidance of a

---

* In the early days of the Soudan occupation it was thought by many that it would be impossible to abolish slavery, and that the employment of forced labour was imperative. As a matter of fact, by the display of a little patience, and without the adoption of any very heroic or sensational measures, slavery has been abolished, no forced labour has been employed, and the country is prospering.

paternal and sympathetic administration, certainly outdistance and leave far behind in the race of supremacy
such systems as those which prevail in San Thomé and
Principe." I trust, and I also believe, that Mr. Harris
will prove to be a true prophet.

It is, moreover, the adoption of the principle to which
I have alluded above which enabled an American Bishop
(p. 109) to characterize as "just marvellous " the way in
which the English are " covering the Continent with
educated natives," and I am particularly glad he was able
to add " with carpenters, bricklayers, and engineers."

In spite, however, of the unstinted praise which Mr.
Harris has to bestow and which makes it clear that,
broadly speaking, we may legitimately be proud of what
our countrymen, both official and non-official, have
accomplished in Western Africa, he indicates certain
defects in the administration, some of which appear to
me to be well worthy of the attention of the responsible
authorities.

In the first place, he says (p. 125) that " between
the British official class and the merchant community a
great gulf is fixed." * If this is the case, there would
certainly appear to be something wrong. There ought
to be no such gulf. But as I presume there is an official
side to the case, which I have never heard, I do not
presume to pronounce any opinion on the merits of the
point at issue. Neither is it altogether pleasant to read
the episode related on p. 151. It is clearly not right to
march into a church whilst service is going on, impound
a number of carriers and " insist on a native clergyman

* See also pp. 94, 95.

carrying a box containing whisky." One may charitably
hope that the facts of the case were not quite accurately
reported to Mr. Harris.

In the absence of adequate local knowledge I cannot
pursue the discussion of this branch of the subject any
further, but there is one observation I should wish to
make. There cannot be a greater mistake than to
employ underpaid officials in the outlying dominions of
the Crown. We do not want the worst, or even the
second best of our race to prosecute the Imperial policy
to which we are wedded as a necessity of our national
existence. The work presents so many difficulties of
various descriptions that if we are to succeed we must
impound into the British service the best elements which
our race can produce, and, as I am well aware, even when
their services are obtained and every care has been
taken, mistakes will sometimes occur in making appoint-
ments. And if we want the best material we must pay
the best price for it. Men of the required type will not
submit to all the privations and discomforts, not to speak
of the dangers of an African career unless they are
adequately remunerated. I know well from bitter
experience the difficulties attendant on paying high
salaries out of an impoverished, and even out of a semi-
bankrupt Treasury. And I also know the criticisms to
which, notably in these democratic days, the payment
of high salaries is exposed. My answer to the first of
these objections is that if the Treasury cannot afford to
give adequate salaries to its European agents it is, on all
grounds, wiser to diminish the amount of European
agency, or even to dispense with it altogether. My own

experience has led me to prefer infinitely the employment of two efficient men on £500 a year to that of four doubtfully efficient men on £250. My answer to the second objection is that those who plead against high salaries are generally very ill-informed of the facts with which they are dealing, and that, if ever there was a case when Government, being better informed, should resist a hasty expression of public opinion, it is this.

Are the British agents employed in subordinate positions in Africa adequately paid ? From all I have heard I have considerable doubts whether they are so. In dealing with this subject, I have heard it sometimes said : " Candidates are plentiful. If we can get a man on £250 a year or less, why should we give him £500 ? " I consider this argument not merely pernicious but ridiculous. It would never be used by any one who has been brought face to face with the difficulties which have actually to be encountered. Its application in practice is liable to lead to very serious consequences in the shape of loss of national credit, and possibly in other and even more serious directions.

Turning to another point, I notice (p. 120) that Mr. Harris states that coloured men are practically debarred from entering the medical service in the West African Colonies, and absolutely in the Gold Coast. If so, I can only say that this regulation contrasts unfavourably with the procedure adopted in other British possessions of which the inhabitants are coloured, and adopted, moreover, without, so far as I am aware, the production of any inconvenience. Possibly there are some special reasons, with which I am unacquainted, which apply to West Africa, but they must be very strong to justify a course so little in harmony with the general practice and policy of the British Government elsewhere.

Mr. Harris deals fully with the subject of education, and in his fifth chapter chivalrously defends the cause of that much-abused individual the " educated native," whose merits and demerits seem to present a striking identity of character whether his residence be on the banks of the Ganges, the Nile, or the Congo. The old complaint with which Indian administrators are so familiar, that the education afforded is too purely literary, re-appears in West Africa. Mr. Harris, however, dwells with justifiable pride (p. 109) on the number of carpenters, bricklayers and other mechanics turned out of the Mission Schools, and (p. 112) he most rightly insists on the importance of extending " that largely neglected branch of education—practical agriculture." He suggests that a Commission should be appointed " to study the whole question of the education of the African peoples in British Equatorial possessions, with the object of ascertaining how far the Government may be able to secure a more even balance between the literary and technical training of natives, and how far it may be possible to so re-adjust existing systems as to avoid denationalization."

My confidence in the results obtained by appointing Royal Commissions is limited, but they afford a useful machinery for classifying facts and sifting evidence, and thus provide some safeguard against the risk, which is nowhere more conspicuous than in dealing with educational subjects, of generalizing from imperfect or incorrect data. Mr. Harris' suggestion on this point will, I trust, receive due consideration.

Mr. Harris also dwells (pp. 113, 114) on the results which ensue when young Africans are sent to England to obtain legal or medical education. " No strong and friendly hand is outstretched to help them, no responsible person comes forward to take them by the hand and bring them in touch with the better elements of our national life. . . . Who can be surprised if the only seeds they carry back to the Colonies are those evil ones which produce a crop of tares to the embarrassment of Governments ? "

If the Colonial Office and the Missionary Societies, acting either independently or in unison with each other, can devise any satisfactory solution of this very important and also extremely difficult problem, they will earn the gratitude of all who are interested in the well-being of our Asiatic and African dominions. Palliatives for the evils which most assuredly arise under the existing system have been tried by the Governments of India and Egypt, but so far as I know the success of these efforts has not been very marked. I may mention that, so convinced was I that the harm done by sending young Egyptians to England for purposes of education more than counterbalanced the advantages which were obtained that at the cost of a good deal of misrepresentation— which was quite natural under the circumstances—I persistently discouraged the practice, and urged that a preferable system was so to improve higher and technical education on the spot as to render the despatch of students to Europe no longer necessary. I fear, however, my efforts in this respect were not altogether successful, for although higher education in Egypt has unquestionably been much improved, the idea that European attainments can best be cultivated in Europe itself has taken so strong a hold both on Egyptian parents and on the Egyptian governing classes, that it is well-nigh impossible to eradicate it.

By far, however, the most interesting and also the most important part of Mr. Harris' work is that in which he deals with the future of the African possessions of Belgium and Portugal respectively. Even if I had at my disposal all the information necessary to a thorough treatment of these questions, it would not be possible to deal adequately with the grave issues raised by Mr. Harris within the limits of the present introduction. I confine myself, therefore, to a very few observations.

As regards the Congo, if I understand Mr. Harris' view correctly, the situation, broadly speaking, is somewhat as follows. Reforms have been executed, and a serious effort, the sincerity of which he does not call in question, has been made to rectify abuses for which neither the Belgian Parliament nor the Belgian nation are in any degree responsible. But although abuses have been checked, the main cause from which they originally sprung has not yet been entirely removed. That cause is that the Government, whose functions should be mainly confined to administration, is still largely interested in commercial enterprises. The State has not yet completely divorced itself from the production of rubber for sale in the European markets. Moreover, the old officials, who are tainted with Leopoldian practices, are still employed. Mr. Harris even goes so far (p. 221) as to state that their presence acts as a deterrent to the employment of Belgian officials of a higher type.* Mr. Harris thinks—and, for my own part, I do not doubt rightly thinks—that so long as this defective system † 'exists, radical reform of the Congo administration will remain incomplete. But radical reform can only be carried out at a very heavy cost, which the Belgian taxpayers, more especially after the assurances which have been given to them, will be unwilling to bear, and possibly incapable of bearing. Mr. Harris, therefore, thinks that the Belgian people will be unable to perform the heavy task which, from no fault of their own, has been thrust upon them. " There are reasons," he says (p. 298), " for believing that the extensive Congo territories are too heavy a responsibility for Belgium."

It is very possible that Mr. Harris' diagnosis is correct. But what is to be the remedy ? The remedy which he suggests is that Germany should take over the greater part of the Belgian and a portion of the French Congo, and (p. 302) should concede " an adequate *quid pro quo* " to France. I will not attempt to discuss fully this suggestion which, to the diplomatic mind, is somewhat startling.

* A good deal, I conceive, depends on the number of old officials whose services are retained, and on the degree of influence they are allowed to exert. I speak under correction, but I can well imagine that the abrupt dismissal of all the experienced administrators, however unsatisfactory they may be in some respects, and the wholesale substitution of well-intentioned, but wholly inexperienced men in their place, might produce inconveniences even more serious than a certain prolongation of abuses in a mitigated and diminished form.

† That this is the main defect of the Congo system has been frequently pointed out both by myself—in a speech in the House of Lords on February 24, 1908—and others.

I will only say that I very greatly doubt the feasibility of arranging any such "adequate *quid pro quo*" for France as Mr. Harris seems to contemplate. The British attitude in connection with any transfer of the Congo State from its present rulers to Germany appears to me, however, to be abundantly clear. If any amicable arrangement could be made by which Germany should enter into possession of the Congo, we may regard it, from the point of view of British interests, without the least shadow of disfavour or jealousy, but—and this point appears to me to be essential—it must be of such a nature as will not in any degree impair the very friendly relations which now fortunately exist between our own country and France. The well-being of the Congo State, however deserving of consideration, must be rated second in importance to the steadfast maintenance of an arrangement fraught with the utmost benefit not merely to France and England, but to the world in general.

Failing any such rather heroic measures as those proposed by Mr. Harris, the only alternative would appear to be to rely on Belgian action, and to exercise continuous but steady and very friendly pressure in the direction of crowning the work of the Congo reformers. It would be unjust not to recognize the great difficulties which a series of untoward events has created for Belgium. It may well be that if this course is adopted the progress of reform will be relatively slow, and that in the end it will be less effective than that which would be secured by an immediate and radical change of system. But I rise from a perusal of Mr. Harris' pages with a feeling that Congo reformers have no cause for despair, albeit their ideals may be impossible of realization in the immediate future.

The case of Portugal is, from the British point of view, if not less difficult, certainly far more simple than that of the Congo. If one-half of what Mr. Harris says is correct —and I see no reason whatever to doubt the accuracy of his facts—two points are abundantly clear. The first is that however it may be disguised by an euphemistic nomenclature, slavery virtually exists in the African possessions of Portugal. The second is that the methods adopted in the repatriation of the slaves are open to very strong and very legitimate criticism. The process of dumping down a number of starving blacks on the coast of the mainland and leaving them to find their own way to their distant homes in Central Africa can scarcely be justified.

Portugal is justly proud of her historical connection with Africa and wishes to retain her African possessions. We may heartily sympathize with this honourable wish. I know of no adequate reasons for supposing that the present political status of those possessions is threatened. But, I venture to think, it would be a mistaken kindness to leave the Portuguese under any delusion on one point. There are some things which no British Government, however powerful otherwise, can undertake to perform. First and foremost amongst those things is the use of the warlike strength of the British Empire to maintain a slave State. In spite of the long-standing friendship between the two countries, in spite of historical associations which are endeared to all Englishmen, and in spite of the apparently unequivocal nature of treaty engagements, it would, I feel assured, be quite impossible, should the African possessions of Portugal be seriously menaced, for British arms to be employed in order to retain them under the uncontrolled possession of Portugal, so long as slavery is permitted. It is earnestly to be hoped that, before any such contingency can arise, the Portuguese Government will have removed the barrier which now exists by totally abolishing a system which is worthy of condemnation alike on economic and on moral grounds.

One further incident in connection with the general question is worthy of notice. Mr. Harris says (p. 200) that a small number of the slaves now employed at San Thomé are British subjects. There ought surely to be no great difficulty in dealing with this class. African experts would probably be able to say whether the claim to British nationality was justified or the reverse. If justified, it seems to me that the British Government should send a ship to San Thomé, embark the men, and, after having landed them at the most convenient ports on the mainland, make suitable arrangements for despatching them to their respective homes.

CROMER.

36, WIMPOLE STREET,
October, 1912.

AUTHOR'S PREFACE

It has become the custom in recent years for writers, particularly those recording their travels in semi-civilized regions, to disclaim in advance any title to literary merit. I do not propose to make any exception to this rule and would plead in lieu of literary style a sincerity of purpose, which I beg my literary critics and superiors to accept. If they feel that the facts and incidents set forth suffer from any lack of literary ability, I can only hope that they will take the information supplied upon some of the existing problems of West Africa and use it in their own skilful way with the object of helping forward the march of progress in West Central Africa.

The information contained in this book is drawn from an experience of West Africa dating back to the year 1898 and in particular during a recent journey of something like 5000 miles through the western Equatorial regions. The principal questions under review, are those which affect in the main the conventional basin of the Congo and the Colonies of the Gulf of Guinea.

It has always been my endeavour to get to know the mind of natives and merchants outside the circle of " the authorities," a habit which I feel has sometimes entailed the appearance of discourtesy, but I know how reticent are the merchant communities, no less than the native tribes, even the most untutored of them, if they see a man or woman holding friendly relations with the powers that be. This method of investigation I have always pursued, with the result that information of the utmost value has frequently been supplied. Whilst, however, I have felt this to be the best course to follow I have, at the same time, tried to place myself in the position of a responsible minister of the Crown, a governor, an official and even a planter, in order that so far as possible I may look at things from their standpoint.

The question may be raised by some of my readers how a man who has spent so many years of his life in distinctly religious work can presume to write upon commercial and political problems. I would make no excuse for so doing, but in justification would say that prior to preparation for missionary work, it is well-known by many of my friends that I held a responsible position in one of the leading commercial houses of the city of London, which, amongst other advantages, gave me a large insight into foreign and colonial questions. My experience of the Congo and cognate questions early brought me into touch with eminent statesmen and well-known public men, including President Roosevelt, Lord Cromer, Sir Edward Grey, Lord Fitzmaurice, the Archbishop of Canterbury, Sir Charles Dilke, Sir Francis Hyde Villiers, Sir Arthur Hardinge, Sir Harry Johnston, Sir Valentine Chirol, Mr. St. Loe Strachey, Dr. Thomas Hodgkin, and my friends Travers Buxton, E. D. Morel and Harold Spender. It is impossible to enjoy frequent discussions with men of such breadth of knowledge, wide experience and high ideals, without considerable profit, and at least

some qualification for a responsible position. If there
is one to whom I am more deeply indebted than another,
it is to Lord Fitzmaurice, whose friendship and counsel
I have been privileged to enjoy in an increasing measure
for nearly twelve years.

My thanks are due to the Editors of *The Times*, *The
Manchester Guardian*, *The Nation*, *The Daily Chronicle*,
The Daily News and Leader, and the *Contemporary
Review* for permission to use material which has already
appeared in their columns. To Mr. Hamel Smith, the
Editor of *Tropical Life*, the Liverpool Chamber of
Commerce, Messrs. John Holt, F. A. Swanzy and Elder
Dempster, for the information and help they have so
kindly supplied to me, and also to my wife for the
assistance rendered to me in the preparation of the
manuscript.

October, 1912.

TO

MY DEVOTED COMPANION

WHO HAS SO PATIENTLY BORNE THE HARDSHIPS

OF TRAVEL AND THE LONG STRAIN OF OUR

LABOURS FOR THE NATIVE RACES

THESE PAGES ARE

DEDICATED

CONTENTS

xxx

CONTENTS

PART III

PART IV

MORAL AND MATERIAL PROGRESS

PART V

LIST OF ILLUSTRATIONS

FOREWORD

WEST AFRICA

WEST AFRICA, as some of us have known her, is rapidly changing. Within the memory of most men, there were deserts uncrossed, forests unexplored, tribes of people unknown. To-day every desert has been traversed; to-day we know not only the forests, but nearly every species of tree they contain; we know, and can locate, almost every African tribe, and almost every foot of territory has passed under the control, for the time being at least, of some alien Power.

At the present moment political boundaries are more or less fixed, but for how long? In Europe certain Powers are, for one reason or another, seeking opportunities somewhere for colonial expansion, and the moment seems opportune for a reshuffle of colonial possessions, but where, and how?

Looking into the Far East, the statesman sees nothing but trouble ahead in the Celestial Empire, to say nothing of Japan standing sentinel over the Orient. South America, with its vast resources and possibilities, might fall an easy prey to an energetic Power, but over every Republic, Monroe casts his protective declaration which, with the march of time, fastens itself ever more firmly upon the vitals of the body politic of the Republican States of South America.

Back to Africa, the searching eye of the statesman returns and rests to-day. There in the Dark Continent are great territories awaiting development, there weak administrations are " muddling along " doing themselves no good, and their neighbours irreparable harm. For those Powers the hand-writing is on the wall ; they must either " get on, or get out," otherwise " like a whirl-wind " some other Power will come and without ceremony bundle them out of the path of progress.

In fifty years the map of Africa will bear little resemblance to that of to-day, but what of the natives ? Are they to have no voice in their destiny ? One listens with impatience to the cool and calculating dis-cussions for a re-arrangement of the map of Africa, which are being carried on without any reference to the native tribes, without any reference to treaty obligations, and with little respect for the fundamental obligations of Christianity, the teaching of which the European Powers claim as their special monopoly.

Commerce, too, is changing ; the kind-hearted mer-chant of West Africa going forth at his own charges, trudging from village to village founding branches, paddling up and down the rivers and planting factories, is disappearing, and the soulless corporation with direc-tors who are mere machines for registering dividends, are taking his place. Commerce in West Africa is rapidly losing all the humanity which was once its driving force.

The natives are abandoning the old forms of warfare. Denied the weapons which would give them equal chances in mortal gage, they are astute enough to refuse to accept mere butchery. They are learning that there are powers mightier than the sword ; education is advancing by leaps and bounds, and the more virile colonies are producing strong men who will make themselves felt before many years have passed over our heads. The African is shaking himself free from the shackles he has worn for so long and is at last beginning to realize his strength. At present Britain, with all her shortcomings, leads the way in giving the native the fullest scope for his abilities. In British and Portuguese Colonies alone in West Africa has the free native the chance of attaining the full stature of a man. In German and French tropical territories, the native is there, not as a citizen, but merely as a necessary adjunct to the production of wealth for the white man. How long he will be content with this position is a question, and evidences of a coming change are everywhere apparent.

Soon the Africa we have known—yea, and loved— will have been hustled away. Its forests, rivers and tribes will possess no more secrets ; gone will be the simple old chief ; gone the primitive village untouched by European ; gone the old witch doctor, and gone too, perhaps, that beautiful faith and trust in the goodness and honesty of the white man—the pity of it all !

Before these changes come, it behoves us to examine closely the great problems before us—the problems of future political divisions, problems of labour, and of education in the largest and fullest sense—and so to readjust our conceptions and laws with an understanding of the natives as save ourselves from repeating the blunders of the past ; blunders which have cost Africa

millions of useful lives ; blunders which have indelibly
stained for time and eternity the escutcheon of Christian
Europe ; blunders for which recompense can never be
adequately made, but which at least should serve as a
warning for the future.

PART I

I

THE AFRICAN "PORTER"

It is almost impossible to exaggerate the part which the African " porter " or carrier, plays in the history of the Dark Continent. The hinterland of the vast tropical regions—a death-trap to every beast of burden—has been opened up by the carrier together with his brother transport worker—the paddler. The heavier burden has, beyond question, been borne by the former, by the countless thousands of hard woolly heads which have sweated under the weight of innumerable bales and cases too often receiving as a reward of their labour an endless stream of abuse. It seemed justifiable to murmur when crossing those swamps and fighting one's way through impenetrable forest, but at a distance, and with time for calm reflection, there can surely be no other thought in the mind of any African traveller than that of admiration, as he pictures those sons of Africa with heavy and cumbersome loads upon their heads, floundering through swamps, or toiling up steep hills and along stony paths, cutting and blistering the feet, while the fierce rays of the tropical sun scorch every living thing. Yet on that carrier goes, footsore, often foodless, yet ever ready to renew the march of to-morrow.

Railways and bridges, steamboats and bungalows,

engines of war, machinery for drilling into the bowels of the earth, lofty windmills, telegraph wires and poles— these and other European conquerors of African air, land and water have by the thousands of tons found themselves hundreds of miles in the interior of Africa owing to the infinite endurance of the African carrier. Abuse him who will, but be sure of one thing, history will yet give him his due.

The railways, bridges and steamboats, would, so we were told, lessen the need for carriers. That they have shortened distances we grant, but so far from the need of the carrier being lessened, economic expansion has increased the demand. The opening up of the country has brought an insatiable civilization into close touch with vast uncultivated tracts of land, with the result that a great impetus has everywhere been given to agricultural development, which in turns calls for an unceasing stream of carriers to feed the railways and steam craft.

Thirty years ago the British colony of the Gold Coast possessed no railways, nor was there any export of cocoa. To-day she exports annually over a million pounds' worth of cocoa-beans, requiring in the season over 100,000 carriers to convey the cocoa harvests to the railways. True statesmanship must always aim at releasing labour from the unproductive task of transport, in order that it may till the soil, but it is doubtful whether the African carrier will ever completely disappear.

Their long procession is never without interest ; every man has some distinguishing mark upon which the white traveller may meditate as he trudges along, now in front,

THE CANOE SINGER.

A LIGHTHEARTED CARRIER.

now in the centre, now again in the rear of a caravan.
What a medley yonder man carries upon his head!
There is the traveller's "chop" box or his bundle of
bedding, to which perhaps is lashed by means of a
piece of forest vine, the sundry goods and chattels of
that simple-hearted carrier—an old salmon tin filled
with odd little packages of salt, chili peppers, bits of
string, possibly a piece of soap, an old knife and the end
of a native candle. There is also the "Sunday best,"
whose owner, while looking happy enough in that strip
of loin cloth held in place by a cheap European strap,
yet strides the firmer and prouder because of that old
cotton shirt and the patched white trousers so carefully
protected by a bundle of forest leaves. Provisions, too,
are there, carefully pounded, cooked and flavoured by
the good wife at home. Those unsavoury manioca
puddings for "her man" are generously accompanied
by her catches of fish, smoked and set aside that he
might each day have an appetizing morsel for his meal.

Other carriers are distinguished by the wounds and
bruises of their calling—one limps along with a sore foot,
but on he goes until the journey's end; others there
are with sore skin or nasty wounds, caused by forest
thorns or rough stones, others whose chafed shoulders of
yesterday now gape and become a resting-place for the
torment of flies; yet, with it all, the impatient traveller
too frequently falls to scolding and even cursing them
for their "laziness"!

No white man should be allowed to travel beyond a
day's journey with a caravan unless he has a few medical
aids for such bruised and wounded helpers, and it will

repay him if human gratitude can be called a reward.
Cuts and wounds are both the inevitable price of African
travel, and it is a necessity and a duty to carry a few
spare bandages and healing ointments.   There is satisfaction too in gathering the sick men round in the evening
and giving them a soothing plaster, ointment or a bandage.
A little human kindness of this nature helps to make the
journey a happier one for all, but alas too often what
the Germans call *tropenkoller* has no conception of
a remedy for complaints beyond the whip or the boot.

The carrier is no more an angel than other human
beings, no matter whether pink or black ; he has all the
imperfections and the love of self-preservation of the
brother who calls himself white.   I remember once having
all the loads laid out ready for the start and then giving
the order for each man to choose his load.   It was evident
the carriers had mentally marked the load each would
like to seize, for a dash was made for a small box only
about 18 inches square and having the appearance of a
20 lb. load—but it was a case of cartridges weighing
80 lbs !   How promptly they all discarded that box and
dashed away for the larger but lighter loads !

Strangely enough the carrier seldom " pilfers " on a
journey.   The white man's goods may suffer depreda-
tions on the steamer or on the train, but on the march
there seems to be a sacred community of interest which
safeguards the goods of most white men as effectively as
if protected by the spirit-haunted herbs and parrot
feathers of the witch doctor, but when civilization, in the
shape of steamers and railway trains, enters barbarous
regions away goes the eighth commandment.

There is one respect in which every African traveller invariably suffers—hungry at the midday hour, he calls for " chop " ; thirsty, he asks for filtered water ; or at night, dead tired, he looks for his folding bed ; he may call in vain, for either from set purpose with some definite object in view, or from stupidity, these essentials to the white man are generally " miles behind."

Probably the carrier is at his best when travelling through the vast forests, where, shielded from the sun by the interlacing trees overhead, it is delightfully cool and the layers of dried leaves render the path as soft and springy as the richest carpet. Carriers and traveller are in high mood and as conversation flows freely the traveller realizes what great students of nature these sons of Africa are. As they walk along, they will name every tree and almost every plant ; they will tell how many moons elapse before the trees begin to bear ; they will give descriptions of edible fruits, the birds and animals which each kind of fruit attracts, varying these running comments by periodic dashes through the undergrowth in search of fruits to illustrate the conversation.

How closely, too, they watch the path for the foot-prints of animal life, never at a fault to identify the prints with their owners, or accurately gauge the time when the creature passed by, begging, if the traces are recent, to be allowed to track the " meat." As time does not concern the hunter, it is generally wise, if there is any reasonable chance of obtaining food for the caravan, to camp for the night. This knowledge of forest life stands the natives in good stead, for not infrequently provisions run out on the long marches and in the absence of human habitations, the question of feeding the caravan becomes a serious matter.

At one time we had marched for days without any opportunity of obtaining a supply of food and the carriers were all suffering from hunger ; in a whole day we seldom found more than a small handful of edible fruit. At last it became almost impossible to push on with the caravan so tired and hungry : I called together a few of the men and asked what we should do, whereupon one made the novel suggestion of " calling the meat." The proposal was readily taken up and three of us pushed on ahead with guns. Arriving at a quiet spot, one of the men—a very son of the forest—fell on his knees, and, placing the tips of two fingers in his nostrils, emitted a series of calls which made that forest glen echo with, as it were, the joyous cries of a troop of monkeys ! How anxiously the tops of trees were watched ! After repeating these tactics in several places in the immediate vicinity for about half-an-hour, a man close to me whispered excitedly " here they come " ! In the distance we could see the tree tops moving, and in a short time a score of monkeys could be seen skipping from tree to tree towards the inimitable monkey cries of our carrier. New life was infused into the whole caravan when they saw the gun bring down four monkeys for the evening meal ; lowering countenances were wreathed with smiles, grumblings and cursings gave place to joyous songs in which even the sick and lame gladly joined. At dinner that night the men were so famished that they could not stop to cook the meat, but contented themselves with merely singeing off the skin and eating the uncooked flesh.

THE VINES OF THE TROPICAL FOREST.

To emerge from the forest is generally to enter once more into habitable country, and there the carriers, no matter how far from home, generally discover a relative —a brother or a sister, a father or a mother. Their relationships are strangely elastic, many an African laying claim to as many mothers as wives, in point of fact the father's brothers and the mother's sisters all rank as the fathers and mothers of the children. The roving British tar may have a wife in every port, but he is surpassed by the African carrier who may have not only a wife but a mother and sometimes a father too in every village!

II

THE PADDLER AND HIS CANOE

CENTRAL AFRICA, the unexplored land of our childhood, is vested with a charm that never ceases to allure, and reveals her deepest secrets only to those who dig deep and risk much to discover them. The rivers with their shifting sandbanks, their treacherous rapids and whirlpools, entice again and again those whom the miasma has threatened to slay, as the rushing current threatens the unwary navigator. The native alone is in any degree immune to the former, and it is he who, with his simple knowledge of the shoals and currents, may venture with his inimitable dug-out where scientific navigation is baffled. Inseparable from the African river is the dug-out, unthinkable are the thousands of miles of navigable waterway without this primitive, though astonishingly effective, craft.

Canoeing in Central Africa may be not unpleasant, providing both canoe and paddlers are amiably inclined. The number of canoes available is so restricted that there is little choice, and comfort aside it is wise always to sacrifice size to reputation, for a canoe with a bad name will dispirit the paddlers. The trimmest and most seasoned craft, capable of holding twenty to twenty-five paddlers, is the traveller's ideal, but the equipment is incomplete without a small pilot boat for surplus baggage, manned with four or five paddlers, who will keep ahead, but always in sight, forewarning of rocks, snags, or sandbanks, and generally discharging the functions of a scout. No less important is the selection of the crew, and these to complete a harmonious group should be volunteers—the best plan being that of getting three or four cheery spirits to select the remainder from amongst their friends.

The African paddler readily responds to an appeal for a co-operative canoe journey, but he dislikes any such undertaking as a mere hired paddler. Make him part and parcel of the journey and a host of potential difficulties vanish. Erect in the bows of the canoe a tiny rush shelter with a bamboo bed for the white man, and only the two final though most important elements remain—music and provisions—the absence of either being equally fatal. For the latter, an ample supply of dried fish and cassava can be stored in the canoe, while a currency in the shape of beads, salt, cloth, pins and needles will do the rest. Without music the African can neither live nor die, nor yet be buried ; walking, riding, eating, digging, paddling or dancing, he must have the rhythm of his music, devoid of charm it may be to the European, but vital to the good spirits of the African. To the accompaniment of an old biscuit tin or a couple of sticks, the gunwale of the canoe, or the leaves of the forest, any or all of which can be made to give forth a sufficiency of barbaric sounds to set in rhythmic motion the voices and bodies of all within range.

With canoe packed, paddlers in position with their

long spear-shaped paddles and musicians with their
instruments, provisions piled high and carefully covered,
the start is made.   Farewells are shouted, and blessings
pronounced which if measured by their volume should
preserve the traveller for all time from hippos and snags,
storms and mosquitos, sickness, accident and even
death.

One sees in the African canoe characteristics as
distinct as those of the paddlers, for with a limited com-
panionship comes a close acquaintance with nature and
things inanimate.   There is the leviathan among native
craft shaped by the chief and his followers from a forest
giant and bearing herself with the proud consciousness
of regal ownership.   In such a craft the passengers need
have no fear for she rides majestically with her bows
reared high, breasting the waves of the tornado-lashed
river or lake, unmoved by the raging of the elements.
She is seen at her best as she glides down stream under
the combined influence of the current and the swinging
impetus of her thirty stout paddlers.   There is the
rickety old canoe with broken stern and crippled sides,
and her leaking bottom stuffed with clay, but there is
life in her yet.   She ships water fore and aft, and amidships too, soaking the traveller's blankets and provisions,
but her long experience gives her an ease in travel which
her younger though stouter relatives cannot rival.
Then there is the lumbering ungainly dug-out, with
crooked nose and knotty sides, unreliable and ill-balanced,
possessing an affinity for every submerged snag.   " Hard
on " she frequently goes and every effort to free her
th reatens to drown the occupants.   Sandbanks she seeks

MRS. HARRIS CANOEING ON THE ARUWIMI, UPPER CONGO.

A RICKETY DUG-OUT.

out too and obstinately refuses to "jump" them. The paddlers will haul her off and curse her roundly for her crooked ways,—but as she was hewn so she will remain. At the other end of the scale is the tiny fishing dug-out of the Niger and the Congo, and their still smaller sister of Batanga in Spanish Guinea, the latter so small that the owner may with ease carry on his shoulder both canoe and fishing tackle, and whilst baiting hooks and catching fish he skilfully sits astride her and paddles with his feet.

Inseparable again from the dug-out is the paddler. Who that travelled with him can forget him? Humorous as the London Jehu of the twentieth century, dexterous as his civilized confrères of the ocean, as adaptable to his surroundings as the clay to the potter's art; at home everywhere and in all conditions in his native land, swimming or standing, sitting or lying, squatting or reclining, sleeping as soundly on the nose of the canoe or the river bank as we in our downiest of feather beds. He is ready and alert with the earliest peep of dawn, as the mists rise from the surface of the river, presenting the appearance of a huge boiling cauldron. Peeping from beneath your mosquito net you see his figure outlined against the dawning light as he keeps a sharp look-out for the hidden snag, and shivers with the clinging chilly mist. His powers of endurance are unequalled, as the rising sun dispels the mists and mounting higher in the heavens becomes increasingly fierce. He still swings his paddle with steady persistence till his body steams with the effort; then after a little halt and refreshment in the friendly shade of the riverside forest, he will go on

until the sun is sinking, and if need be, still on in the
moonlight, singing his monotonous boat song, occasionally
varied by a running commentary from the leader on the
incidents of the journey, the peculiarities of a certain
paddler, or the ways of the white occupants of the canoe.

During the whole day long the paddler will pursue
his task, I see him now almost unconsciously bending
his body with each dip of the paddle, till a sudden slowing
down followed by a profound stillness arrests the attention. I can again hear those whispered voices as the
gentle lapping of the water against the canoe side ceases,
and the boat is still. A monkey has perhaps been seen
overhead springing from bough to bough, or sitting
nibbling the fruit of some forest tree, or it may be an
edible bird with flesh as tough as its plumage is gorgeous,
that watches us till the gun booms out and the creature
is brought down. For a moment it struggles in the river,
then with a sudden splash, a man is swimming with
powerful strokes towards the prey which he a moment
later lands in the canoe, while the rest look approvingly
on at their prospective meal. With spirits heartened,
on they go, singing of their capture and the feast which
is to follow, till turning a bend in the river the destination is at last in sight.

And how they love a race! Let them but see a
competitor ahead bound for the same goal, and despite
their long day's paddle they will redouble their strokes.
Caution is thrown to the winds and the canoe springs
in a mad gallop, rocking to and fro, pitching and tossing
against the current until the rival ahead, scenting a race,
enters the competition with keen zest. At such times I

have found all warnings are in vain. With a rapid girding up of the loin cloths as the boats proceed, a re-arranging of the cargo, children, dogs, fowls, baggage and all—the race begins in real earnest. With much shouting and good-natured banter the one or the other will take advantage of every prospect of an up-current ; now out again in midstream to avoid a snag ; a paddle breaks in the effort, but is quickly discarded and another seized without lessening the speed—and on they go, each determined to win or sink their rival. The boats ship water, but are made to right themselves with marvellous ingenuity and then both stop to bale out, while the paddlers exchange good-humoured threats, gibes, curses and defiance.

On again they go with little advantage to either side, and the word is passed for the " master stroke." Madder than ever is the race ; the white man may shout but they pay no heed, for young manhood has lost all sense of danger. At last the opportunity occurs for the final advantage for the river must be crossed at yonder point. Often have I tried to avoid this danger, proceeding first to command, then to plead, but in vain. I might as effectively have tried to control a hurricane with a feather ! To clear the point with its snags, one canoe must fall behind or cross the rival's bows—to give up and fall back is impossible. The attempt is generally made by the smaller boat to cross the bows of her more powerful rival and though occasionally successful she is more often struck amidships and disappears completely— canoe, paddlers and all !

A great shout goes up and the victors splash in to

the rescue, seizing the mats, baskets, provisions and sundries, which float off in every direction. The crew, as much at home in the water as on land, come up one by one and others dive to seize the stern of the sunken canoe. With vigorous pulling and pushing, the water is swished out till she floats again, and in the vanquished spring, again baling out the remaining water with their feet, till it is once more fit for occupation, and every one is prepared for the last lap of the journey. The men take their beating well, enjoying the laugh against themselves. That night all sleep together in a friendly fishing encampment, while the white man curls up in his canoe, and listens to the merry paddlers as they recount with evident enjoyment the story of their five-mile race.

Who that has found a home and nightly shelter in an African canoe will not, as he quits it after many days, feel that he is leaving an old acquaintance behind him. Through the twenty-four hours of sleeping and waking, the canoe and the traveller have adapted themselves to each other's limitations, and the recently vacated canoe speaks as eloquently of emptiness as the vacant chair.

III

THE AFRICAN FOREST

THERE can hardly be any experience more exquisitely luxurious than that of wandering on through the primeval forests of Central Africa. The traveller whose daily round confines him to the great cities of a hustling civilization finds himself in perfect solitude, perhaps for the first time in his life. Every step he takes brings before him some new wonder in nature's garden; every hour in the day is alive with fresh experiences.

Surely there is no language which aptly befits the transcendent beauty of nature awaking to greet the new-born day. During the night, giant forms have roamed at will through the silent glades and recesses of the forests, but with the peep of day they have retired to their lair. Those feathered sentinels, whose hoarse cry rings through the night hours, have perforce veiled their eyes at the awakening of their comrades who strike the sweeter chords befitting the glad hours of day. Throughout the night the trees made monotonous music by the incessant drip drip of their tears, but with the morning, the warm sun has bidden those tears begone.

When daylight breaks through the tree tops, the boughs sway here and there as the monkeys, springing from tree to tree, gambol with their fellows, only ceasing

for a momentary peep at the strange intruders of their
sylvan preserve, as the undergrowth crackles beneath
the travellers' feet and the squirrels dart across the
pathway seeking a safer retreat.   The sight of the white
clad figure, moving rapidly through the mass of undergrowth, startles the mother bird from her nest, and off
she goes shrieking for her mate and warning her fellows.
Yet over all, a silence broods and the traveller falls to
constant musing as he wends his way.

For miles the dense forests will shut out the sun, and
then perhaps where a lofty giant tree has fallen in decay,
a slanting ray of sun will gleam through the leafy roof
turning the pathway into a smiling track of iridescent
moss and fern.   A few yards further and the path
descends abruptly into a woodland stream, bridged by
rustic logs, only possible of fording in mid-current by
creeping warily along the trunk of a tree which some
thoughtful passer-by has felled. The logs and trees
which lie rotting in all directions are the home of shimmer-
ing mosses and tiny fern. Beside the slippery and
tottering causeway there shines many a filigree globe of
purest opal and cunning design like a fairy's incandescent
light guiding the steps of the unwary traveller.   They
are but insects' nests, as fragile as delightsome, crumbling
at the touch.

The traveller can never proceed many miles on his
journey without meeting the dark lines of driver ants.
At a distance of fifty yards, all one sees is a uniform
brown line, sometimes two inches wide, sometimes as
many feet. Drawing nearer they are seen to be well-
ordered regiments, thousands strong, with scouts, baggage

WILD FOREST FRUIT.

bearers, captains and field-marshals. Their enemies have fled at their approach and they are masters of the field. On they will march, in faultless array, their countless thousands obediently passing at a double their immovable field-marshals, and as they proceed every living thing flees before them. Now perhaps they disappear through some subterranean passage, tunnelled out by their indomitable energy, to reappear on the surface when it suits their plan. You may scatter them if you dare, but you can daunt them never. Sweep them into heaps, kill them by the hundred, burn them by the thousand, and tens of thousands surge forward, to fill up the ranks, and remove the dead. Then the regiments will grimly move once more on their way—an incentive to higher organisms.

The African forests teem with life, for the most part silent ; even the great beasts glide along in perfect quietude—not till you are upon them do you realize the proximity of the elephants ; then, unless the traveller be a Nimrod, his greatest concern is to avoid a possible encounter. They love most a quiet glen near the forest stream, where they will plough the earth in all directions and everywhere leave the impress of their giant limbs stretched in gymnastics all their own ; here and there scattered over their playground lie scores of trees athwart each other, evidence of no woodman's axe, but of the entwining grip of the monster's trunk, who in his unrivalled strength delights thus to shew his power in his own domain. Everywhere, too, the great forest apples lie idle after their sport, and the natives tell how they spend hours hurling these great balls at their fellows.

The rivers which everywhere feed the African forest, coupled with the tropical sun, give luxuriance to all nature. Vines are there as much as three feet in circumference, moss-grown and gnarled with age, born perhaps when the parent acorns of our oldest oaks were yet unformed. Sometimes like huge serpents they coil themselves in a tortuous grip round two or three trees, each of which may be ten times their own size.

There is beauty too in these silent forests, when at intervals on the march the traveller, almost unconsciously at first, begins to inhale the fragrant odour of some delicious perfume sent forth by modest blooms that shun the gaze of man. A little searching beneath the undergrowth, or in the tree tops overhead, reveals a bloom upon which the eye gladly lingers—trails of waxen jasmine hanging from the bush in exquisite profusion.

There is beauty, too, in the forest decay, in the fallen tree trunk, whose rotting bark and ugly torn stump are transformed by tufts of gracefully drooping fern, while tiny rootlets smile from out every crevice. There is beauty, too, in the fungus growths, tinted and white, or the perfection of coral, or blooms whose purple depths suggest some cherished hot-house flower.

The experienced traveller is quick to note signs of a change ; the pathway leads uphill and the absence of giant tree trunks denotes that he is treading a once cleared and populated region. That hill, whose summit is capped with foliage, was once a village landmark, beneath it, myriad termites live and pursue their daily toils through tunnels and chambers that they have shaped by their

WILD FOREST FRUIT.

THE "ELEPHANT EAR" IN THE WET SEASON.

countless thousands. Were man's three-score years and ten twice told devoted to the study of the ways and purposes of the unheeded occupants of our earth, he had but then begun to learn the alphabet of nature's infinite resources.

Close to the termite hills the half-buried foundations of primitive dwellings speak of departed life, and in the Congo, hundreds, yea thousands, of these mark the spot where once the children of nature lived out their simple life, till civilization strode through the land treading ruthlessly down the souls of men. They have gone and their haunts lie deserted, but their monuments remain. The discarded kernels of the housewives' palm nuts have taken root and now rear their graceful fronds on faultless trunks like capitals of Corinthian pillars in some cathedral aisle. As if by design they ranged themselves thus. In these silent groves the traveller treads reverently upon the grassy floor ; no monk is here ; there is no echo of the choristers' song, but nature has reared her temple where myriad voices rejoice and sing their song of praise, unfettered by the forms and creeds of man.

The long day's tramp is now over ; the sun is setting and the birds are carolling their evening song, as the traveller emerges into the open space beside the gleaming river, flowing swiftly onwards with its errand to the sea. The glow of the departing sun tints the clouds with purple and gold outshining in glory the loveliness of the morning. Surely the heavenly regions are not far beyond, and this is a glimpse behind the veil. The afterglow has departed and the world of man falls asleep till the twittering of the birds heralds the approach of another day with another march through the inexhaustible forests of tropical Africa, where verily

> " Earth is crammed with Heaven
> And every common bush ablaze with God."

IV

A MEDLEY OF CUSTOMS

A LIFETIME spent amongst a single African tribe would scarcely exhaust its folklore and customs. Awaiting scientific investigation there is throughout the African continent a wealth of lore and superstition.

To him who would discover the hidden life of the African infinite patience is essential. It is useless to force information ; the best plan is to wait until the " spirit moves " the old woman or chief to tell you something of the inner life of the tribe. Perhaps the time and conditions which most contribute to a flow of talk are a moonlight evening around the log fires and cooking pots.

I see them now—these simple Africans, seated around the great earthenware pot awaiting the meal of boiled cassava, pounded leaves or steamed Indian corn. I hear that grey-headed old chief, with low musical voice, passing on the traditions of past generations, so " that the boys may know something of the early history of their race." All the old stories familiar to civilization are there. They all know that " man first went wrong through woman gathering fruit in the forest," the only variation is that the kind of fruit differs in different parts of West Africa, but it is always a forest fruit, always the woman tempted the man; always man succumbed! Then the old chief will turn to the oft-told story—the sacrificial efficacy of the young kid. It is remarkable how closely this custom resembles even to-day that institution of the Pentateuch. The young kid must be free from all disease, a perfect animal in every respect. When killed the blood is carefully sprinkled on the lintel and on each door-post. Other familiar sacred institutions are passed under review. Then the animal kingdom comes under discussion, and the whole series of Uncle Remus, with but slight variations, secures the rapt attention of the listeners. It is at such times as these that the student gets beneath the surface of polygamy, burial and marriage dances, cicatrization and the more serious subjects of land tenure, tribal laws, social ties and domestic slavery.

Not all tribes are equally interesting, probably the Baketi tribes on the upper reaches of the Kasai river provide the greatest wealth of interesting customs and folklore. Their grotesque images, carved in wood, grin at the traveller from the door-posts of the houses, and passing through the villages one has to be extremely careful not to tread upon one of the fetishes which are scattered along the walks in great profusion. One day I saw three separate fetishes within a single square yard, and these, the father explained to us in his simple way, he had purchased at, to him, a heavy cost, hoping thereby to restore to health his only daughter. Not only does the Baketi fill his town with fetishes and wooden images, but in the forests which separate village from village, almost every tree along the pathway has rudely carved

THE " HEALING " FETISH.

THE BAKETI MEMORIAL GROUND.   TREES UPROOTED AND PLANTED
BRANCHES DOWNWARDS IN MEMORY OF THE DEAD.

on its trunk the grinning face of some impossible human being.

The Baketi, too, is probably unique in his memorial grounds. Most African tribes bury the dead in the heart of the forest, but at the same time near the village a memorial ground is set apart on which are erected tiny memorial huts, which the restless spirits of the departed may inhabit if they so choose. There, when the spirit pays such visits—as all good spirits do nightly—he finds his loin cloth ready, the spoon with which he ate his food, the bottle from which he drank, his battle axe and cross bow which played havoc in many an affray ; there is generally too a spread of Indian corn or other food, which the thoughtful and sorrowing wives have placed in readiness for his return visit to earth. How safe these memorial tombs are from desecration may be gathered from the fact that very frequently considerable sums of native currency are strewn upon the floor. These little tombs are also surrounded with numerous carved images erected on poles. The Baketi have another custom which is, I believe, quite unique in West Central Africa. Outside every village are large forest clearings covered with grass, and dotted over these meadow-like lands may be seen the strange sight of trees rooted up and planted upside down—the branches having been lopped off or the tree trunk cut through the middle and planted with the roots in the air. The sight of these clearings, involving a considerable expenditure of labour, covered with scores—sometimes hundreds—of these symbolic monuments, is most impressive.

The Baketi have elaborate ceremonials at births and

marriages. A special house is always built for the birth of a child, the mother being conveyed to the dwelling an hour or so before the expected time, as is likewise the case with a dying person. Another curious custom which prevails amongst these people, and strangely enough we found precisely the same custom a thousand miles north amongst the Ngombe tribes of Bopoto, yet nowhere in the intervening territories, forbids any young woman to definitely enter into marriage relations until one end of the interior of her house is closely packed with neatly cut logs of firewood! This usually means about three hundred logs, measuring eighteen inches in length and two feet in circumference. The idea appears to be that of demonstrating the domestic capacity of the bride-elect.

With every West African tribe there are customs peculiar to the individual community, but they are generally trivial, or variations of customs prevailing amongst the surrounding tribes. Amongst Congo tribes only the Baketi apparently possess customs so completely unique.

(a) CICATRIZATION

Cicatrizing is practised more or less over the whole of West Central Africa. In some parts like the Bangalla and Equatorial regions of the Congo, the patterns are extremely elaborate and involve much patient labour on the part of the artist and prolonged suffering by the individual.

Cicatrizing is often confounded with tattooing, but the latter process is entirely different, and is of course

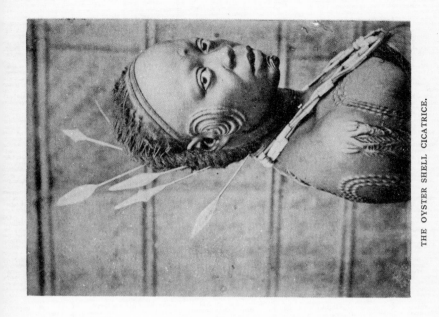

THE OYSTER SHELL CICATRICE.

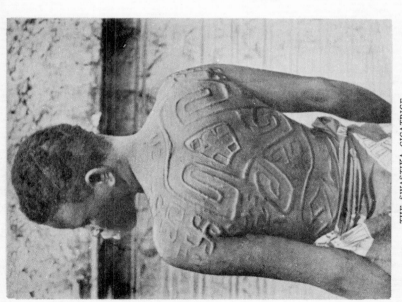

THE SWASTIKA CICATRICE.

most largely in vogue amongst the Maoris and seafaring men. The word cicatrization is derived from the French medical term which designates the scars left by a healed wound and implies a raised portion of the flesh, whereas tattooing is an indentation coupled with the insertion of indelible dyes. Strangely enough the Baluba tribes south of the Congo tattoo themselves, and in this respect are unique in West Africa. Both men and women readily subject themselves to the cicatrizing knife, but generally speaking women are more liberally marked than men.

In the Bangalla regions of the Congo, the facial markings resemble the surface of a coarse rasp, whilst the women content themselves with large shell patterns on the lower part of the stomach. Along the main Congo and some of the tributaries, the marking which finds most favour is the " coxcomb " in the centre of the forehead ; this is sometimes cut quite deeply. The hinterland tribes of the Equatorial rivers almost without exception adopt the oyster shell pattern just below the temple, but the women, in addition, are prodigally marked with " knobs," small " oyster shells " and " bead strings " all over the body, particularly on the thighs. Amongst the Batetela, the forearm is usually covered with a pattern identical with the Cornish " one and all " motto, often also with a sunflower pattern running from the navel up to the shoulder, sometimes to the right, but more often to the left. In the Kasai territories there is first the one general cicatrice imposed on the people by the historic northern conqueror Wuta, a " white " chieftain of prodigious valour and energy, who, apparently more than five hundred years ago, swept through the whole region founding new dynasties and placing the tribes under tribute of soldiers and money. This hustling personage, it is said, reached what is now Rhodesia, but so great was, and is, the fear of his spirit that everyone to-day bears his cicatrice. The Bakuba, Bashilele, Baketi, Bushongos and Lulua, all bear their distinctive marks, many of the women having the whole thigh covered with a " herring bone," and the men carrying a mark similar to the Grecian " key " pattern. In the Portuguese Enclave and the Mayumbe territory of the Congo, the whole of the back is frequently covered by a single pattern and on the back of one woman we found a marking which is clearly the Swastika.

The operation is, of course, distinctly painful. The subject sits on the ground or on a log of wood, whilst the operator cuts deeply into the flesh with the knife held at such an angle that a considerable wound will result. Think of sitting still whilst this crude hand-made piece of native steel is dug into the flesh something like twenty or thirty times within half an hour ! Once I was able to watch the process ; the woman desired a " lace pattern " made from the shoulder blades to the waist, involving altogether four lines, which meant nearly two hundred cuts. She sat outside her hut, and bending down slightly to stretch the skin, the intended pattern was marked in chalk, and then the operator, taking his small cicatrizing knife in his right hand, proceeded to grasp between the thumb and forefinger of the left successive small portions of flesh, gashing each till the blood flowed freely. Then he started the other side of the body, returning again to cut the third line, and

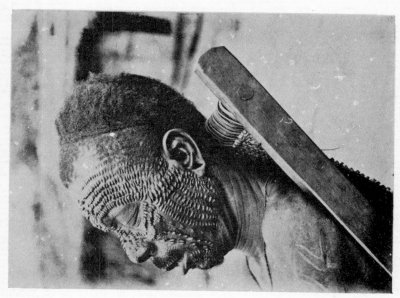

THE BANGALLA "RASP" CICATRICE.

CICATRICED WOMEN OF EQUATORVILLE.

back to the second to link the pattern up with the fourth.

I watched the woman closely, and as the knife dipped into the flesh she made a grimace, but between the cuts, laughingly and with considerable spirit replied to my comments. At the conclusion of the operation, she calmly walked to the nearest tree and gathered a few leaves to wipe up the blood which by this time was streaming down her body. The operator, according to custom, threw over the wounds a handful of powdered camwood which, however, has less antiseptic than drying properties.

It is not easy to light upon such operations, which are generally carried out more or less privately, and in all my years of residence in Africa, this was the only occasion on which I have been able to watch throughout an elaborate cicatrization. It is, however, a familiar sight to meet natives with their bodies newly cut. On the day after the incisions have been made the wounds swell and suppurate, greatly to the delight of the hosts of insect life which swarm everywhere in Central Africa. These surround the wounded body of the native and only by a continuous flicking of grass or twig brushes can the suffering victim obtain even comparative freedom from the tortures which every movement of the body imposes, but in the course of a few months the pattern originally cut in the body stands out firm and clear. In those cases where still more emphatic designs are desired, the cicatrice will be re-opened and raised higher still until the prominence is quite pronounced, in others, after a lapse of a few months, still more lines and still more " knobs " will be added until the age of twenty to thirty.

After this the desire for adornment ceases and the body rests from its tortures.

What is it that attracts ?   What power is it which buoys up the spirit under these painful operations ? What is the secret which gives this insatiable desire for fleshy adornment ?—a desire firmly rooted in the breast of every section of the community and shared by young and old alike.   I well remember an orphan child, of about three summers, standing in the roadway crying bitterly, and upon my asking the cause, she told me that being an orphan no one had enough interest in her to cut a " coxcomb " on her forehead.   Secreting a small bottle of red ink, I told her to sit on the table, and by a series of pinchings and finger-nail marks on her forehead, coupled with a smearing of red ink over my white hands, calmed the little mite into the belief that her heart's desire was being gratified.   After about ten minutes she was supremely happy in the thought that she too possessed a " coxcomb."   Her delight was unbounded, until the little mischief caught sight of her natural forehead in a mirror !

No doubt the principal motive for this passion is the love of personal adornment, of which the African assuredly does not retain a monopoly.   Hitherto the hinterland tribes have had no access to those artificial aids to personal adornment, which are laid so temptingly before the youth of civilization.   They will tell you they have had no alternative but to " adorn " their only garb—nature's dusky skin, and none would deny, that there is a certain beauty even in these barbarous forms of embellishment. The critic may observe that the beauty of womanhood is obviously not enhanced by the bold use of the cicatrizing knife, but I would remind that critic that the wife without a body fairly well covered with cicatrization finds but scant favour with the other sex.   In Africa the European youths of fashion have their counterpart, and in the direction of the most daintily cicatrized maiden, are cast the most amorous glances, and offers of handsome dowries to the admiring parents for the hand of their captivating daughter.

Other reasons doubtless play a part, among them the question of tribal ownership of wives, and the necessity of placing a distinctive and indelible mark upon the body.   Constant internecine warfare, too, demanded a mark which would make easy the task of discriminating the warriors of the respective combatants.

Patriotism, relationship and love of adornment, combine in giving to the African the extraordinary fortitude which this prolonged operation demands, but the disappearance of internal warfare, the increasing importation of cheap jewellery and gaudy clothing, and the advance of Christian civilization, is robbing this custom of its *raison d'être*, and in another generation the little African boys and girls will only learn from books of this curious custom of their grandfathers and grandmothers, for cicatrization, as practised to-day, will have perished within another twenty-five years.

(b) PERSONAL ADORNMENT

Left to nature, the African, dissatisfied with his personal charms, looks about him for some means for adding adornment to his body.   In the absence of finely woven

cloths and silks, he covers his person with ornamental
markings, and his woolly hair he makes to take the place
of head-gear. In two respects only his tastes accord
with those of the European—metal ornaments and
rouge powder.

Most African tribes wear some cloth. The wild
Ngombe on the southern banks of the main Congo, skilled
in ironwork but ignorant of weaving, wear a vegetable
cloth which they strip from the inner side of the coarse
bark of a forest tree. Many of their women content
themselves with only a few cicatrized patterns, and this is
most noticeable in the hinterland of Bangalla, north of
the Congo. A peculiar feature, however, is that all these
women, though completely nude, wear a thin piece of
string round the loins. When photographing a group,
I suggested the removal of these strings, because they
seemed to imply that normally a cloth or leaf was thereby
suspended; but the women, at this, to me, most innocent
suggestion, all became exceedingly angry and threatened
to run away. Finally, I managed to restore good relations,
and we succeeded in obtaining an excellent photograph.
It was evident that some deep significance attached to
wearing this almost invisible cord, but what that significance was I could not discover.

Hairdressing ranks almost equal in importance with
cicatrization, and practically any day the traveller
passing through the villages may see some native stretched
lazily upon a mat on the ground, the head resting on the
lap of the hairdresser—generally one of the opposite sex.
In Spanish Guinea, and on the islands off Batanga, the
style of hairdressing is that of long plaits, sometimes a

BANGALLA BABE WITH HEAD TIGHTLY BOUND.

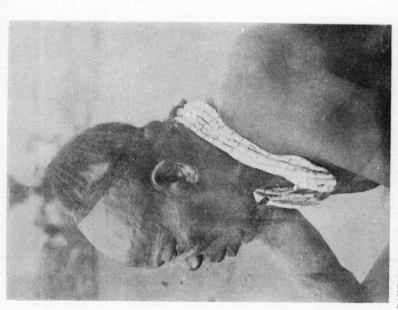

BANGALLA CHIEF WITH HEAD TIGHTLY BOUND FROM BIRTH.

dozen in number, running out in all directions from the top of the head. In French and Belgian Congo the style most favoured is the helmet and in some cases the mitre form ; in these the hair is braided up until it adds apparently about five or six inches to the stature. In many parts of the Cameroons, as well as in French and Belgian Congo, the hair thus built up is covered with a mixture of oil and camwood powder, and thus offers a solid protection against the fierce rays of the tropical sun.

Amongst the Boela people of Bangalla, the custom prevails of binding the crown of an infant's head with tough cord soon after birth, and this head-binding is maintained throughout life. The effect is that of an elongated or sugar-loaf skull which is greatly emphasized when the hair is prominently braided around it. We observed men of all ages with their heads bound in this manner, but they did not appear to suffer any discomfort, and the mental powers of the tribe were in no sense below the average.

Rouge finds great favour in the personal adornment of the African. The powder is obtained from the camwood tree, and in almost every well-regulated household in the forest regions may be seen let into the ground a log of wood some eighteen inches in diameter, while a piece of smaller dimensions lies near at hand. The housewife, in order to obtain the colouring, rubs—or more correctly grinds—one piece on the other, which, with the aid of either water or oil, causes a thick red paste to exude, which is then made into cones and placed in the sun. When thoroughly dry, it is either pressed into a powder and sprinkled over the body, or the person is anointed with a mixture of the powder and palm oil ; in either case imparting a bright red appearance.

In war times, at festivals, and on feast days, an enormous amount of rouge is used, and the red bodies of the tribes are rendered extremely grotesque by the addition of white clay markings which stand out very clearly on the red background.

For the most part the West African tribes extract all the hair from the body with the exception of the head, the beard and moustache. The task is almost a daily one, and in the case of a man is generally undertaken by one or more of his wives. Little boys and girls submit willingly to the removal of their eyebrows and eyelashes.

Brass anklets and necklaces are much prized by the natives throughout West Africa. The Mongo tribes of the Congo wear anklets weighing sometimes 10 pounds on each ankle, and the whole set of ornaments, including the collar, will turn the scale at 35 pounds. In the Leopoldian régime these valuable ornaments were a contributory cause to the atrocities, for the rubber soldiery would always seek out the women in possession of such anklets and collars, and, as they were welded on the body, would not hesitate to chop off the foot, the hand, or even the head in order to obtain the ornaments.

I once heard a neat retort from an African woman. The questioner was a white lady who had been pointing out the pain caused by wearing these heavy articles of adornment. The dialogue ran as follows :—

White Woman : Why do you wear anklets which cause you so much pain ?

A FIVE FOOT BEARD.

STYLES OF ARUWIMI HEAD-DRESS.

African Woman : Beauty is worth pain.

White Woman : Surely you do not suffer such torture in order to appear beautiful ?

African Woman : Tell me then, white woman, why do you suffer pain by tying yourself so tightly in the waist, like a woman suffering the pangs of hunger ?

How far these simple customs should be checked has always seemed to me a matter of doubt, but in the internal government of missions they cause serious dissensions among the staff. Not a few missionaries, and some government officials, seem to feel called upon to place these old-time customs almost on the level of criminal offences.

In one mission no natives may sit down to Holy Communion with their hair braided and oiled, nor may they enjoy the full privileges of Church membership if they use camwood powder on their bodies ; this is the more outrageous when, within a few days' canoe journey, there is another Christian mission where one lady missionary at least is evidently well acquainted with the use of delicately scented rouge. In another mission, cicatrizing, the extraction of the eyelashes, men dressing the hair of women or vice versâ, are sufficient to warrant suspension from Church membership.

In all conscience there is enough that is evil in humanity, both white and coloured, to make the decalogue sufficiently hard of attainment, without human agencies arbitrarily introducing non-essentials which make it grievous to be borne.

(c) "THE ANGEL OF DEATH"

The wildness of the African hinterland, the frequency of bloody feuds, the ever present unhealthiness, almost daily materializes the hand of death. From the moment the traveller touches the coast of Sierra Leone, he is never far from the tragedy of early and violent deaths, accounts of which reach him at every port.

The native's fear of death is immortalized in his many boat songs, his legends and traditions, as well as in those elaborate systems of fetishism which are used to ward off the imaginary proximity of Death's angel.

This was the feature of African life which so impressed Du Chaillu on his first visit to West Africa. "Are you ready for death?" he sometimes asked the natives. "No," would be the hasty reply, "never speak of that," and then, says Du Chaillu, "a dark cloud settled on the poor fellow's face; in his sleep that night he had horrid dreams, and for a few days he was suspicious of all about him, fearing for his poor life lest it should be attacked by a wizard."

Cursing in West Africa, which almost invariably takes the form of invoking death upon some relative, is one of the most frequent causes of trouble. A curse hurled at himself, the African merely resents, and returns the compliment, but let a man invoke death upon another's mother or sister, and the dagger leaps instantly from its scabbard, or the spear goes hurtling through the air with deadly precision.

"May you die" is the most common form of cursing, which brings the sharp retort, "And you also." The curses, " May the leopard catch your mother," " May the crocodile eat your sister," call forth instant battle. The explanation of this strong resentment and intensity of feeling is found in the fact that the African firmly believes that when a curse is pronounced the unfortunate person is thereby accursed.

No man ever goes on a journey, no matter how short, without a string of charms about his neck, to ward off the grim form of death, which he believes lurks in every forest, along every river, in every home. There is one charm to protect from violent death through wild animals, there is one to protect from death at the hands of strangers, but chiefest of all is that little charm stuffed away in the ram's horn, which is a perfect safeguard against the death curse of strangers whom the traveller may meet when on his way from village to village.

The traveller cannot escape the sorrow and despair of death which surely is nowhere so marked as at the death of the African. For days, maybe, the sufferer has lain without any perceptible change, either for better or worse ; then, perhaps, the watcher observes a sign which shews that the end is not far off, and the word goes round the village that Bomolo cannot live long.

Silently, one after another, the relatives creep into the hut and sit upon cooking pots, mats, stools and logs of wood, until the hut is filled with men and women knit together with a common sorrow. The strong man they have remembered in the sylvan chase, the keen fisherman, or possibly the courageous warrior they have known and admired, and in their beautiful simplicity loved, is stretched upon the hard bamboo bed which his busy hands had made. The watchers can see that it is only a matter of hours and the general weeping is at first silent, occasionally ceasing when the sick one speaks or calls for something. The nearer relatives rub and bathe the limbs which begin to chill ; one or two affectionately hold a foot, a hand, or a finger ; the favourite wife, as her right and duty, tenderly nurses the head.

In proportion as the weakness increases, the crying becomes more audible ; then louder still the women cry, invoking all the spirits of the other world to surrender their grip and restore to life and vigour their beloved tribesman. Some momentarily cease crying and call to Bomolo to " speak words of farewell," and the fact that the dying man is unable to reply is a signal for louder wailing still. At last comes the dreadful moment when their friend ceases to breathe. For the space of a few seconds, a breathless and awful silence prevails, whilst brother and wife listen to the heart beat ; then, with a terrible shriek which rends the air, the wife cries, " He is gone ! "

Words fail to describe this scene ! How can the pen adequately portray the bursting of the pent-up misery of these scores of relatives as, in their agony, they twist and writhe in the dust. Wildly despairing, they grasp in frenzy the corpse or the bed, and then releasing their hold, they throw up their arms and again roll in the dust, not infrequently into the log fire which smoulders on the floor of the hut, scattering the embers amongst the tumbling and twisting mass of wailing humanity. What matter those burning scars ?—the frenzy of a terrible

THE WITCH.

SLAVE GRAVEYARD ON THE ISLAND OF SAN THOMÉ.

sorrow consumes reason and chases into oblivion the pains of cut, bruised, scalded and burnt bodies.

An hour later, the storm having spent its fury, the body is washed and prepared for the grave, but the wailing still goes on rising and falling in a monotonous cadence like the moan of a dying gale at sea. There is no escape from that never-ceasing death wail until the body is buried, which, in most villages, is generally within forty-eight hours. Then the tide of weeping turns. A reaction sets in and the weird dancing to drive away the evil spirits continues throughout the night, until mourners and relatives revive sufficiently for the task of partitioning the wives and other worldly goods of the deceased.

The death customs differ with almost every tribe. In the watershed of the Lopori, Aruwimi and Maringa rivers of the Congo towards the Egyptian and Uganda borders, the corpse is frequently hung for weeks over a fire and thoroughly smoke dried. A similar custom prevails in certain parts of the middle and lower Congo. The corpse, however, is dressed in the best clothes and placed for a day or two in a life-like sitting posture—a gruesome and unnerving sight for the passing European. A hut in which a traveller was resting on his journey was seen to have suspended from the roof a deep wicker basket, from which a dark round object protruded. This, on inquiry, he found to be the head of a child whose body, after being smoke-dried, was hung there by the mother that she might look upon the features of her cherished infant. Amongst the Bakwala tribe, the custom prevails of smoking the body of a deceased wife who may be the daughter of a distant tribe, in order that she may be sent home and find burial amongst her own people.

Some of the Bakuba tribes on the Kasai, before life is actually extinct, seize the body, bundle it unceremoniously out of the hut, and then raising it shoulder high rush off to a distant and unoccupied hut that the spirit may there take flight, and not from the home which they believe the spirit would henceforward haunt. It is there prepared for burial, the whole village meanwhile gathering at the house of the deceased to take part in the general wailing.

(d) PEACE AND ARBITRATION

Most African tribes set the civilized world an example in their unwritten methods of preventing war, or, after war has been declared, of bringing it to an early termination. If it were possible to exile the Foreign Ministers of the Great Powers of Europe to the hinterland of their respective colonies—Sir Edward Grey to remote Barotseland, Baron von Kilderlen Waechter to the Sanga in German Cameroons, and Monsieur De Sélves to the Ubangi—where they could divide their time between fishing and studying the peace principles of barbarous tribes, I have little doubt they would return to civilization with more practical ideas upon peace than they will ever learn in the despatch encrusted offices of London, Berlin and Paris.

The African detests war and will make great sacrifices to prevent the outbreak of hostilities. The two principal causes of war are (1) land; (2) wives. Slave raiding does not belong to the African; the Arab imported it. Before war breaks out there is first the " palaver," which

THE WITCH DOCTOR WITH HIS CHARMS FOR EVERY ILL.

may last many days or weeks. In palaver the debates differ but little from the parliaments of the world, except perhaps that custom keeps womanhood out of general debates, although where the particular interests of women are concerned, I have seen them throw themselves into the debates in a manner no whit less collected and impressive than the men.

The African revels in debate, and possibly this accounts to some extent for the admitted passion for litigation which now animates the civilized centres of the African colonies. The orators of the primitive tribes are no less masters of the art than their eloquent compeers at Lagos and Freetown. I was once asked to visit a first-class palaver and found a huge semi-circle of people closely massed together. Soon after my arrival the chief took his seat and one could almost hear the police-men of St. Stephen's calling out, " Speaker in the chair ! " for a similar signal was given for the palaver to commence.

The chief, surrounded by his advisers, called upon the speakers in turn ; first to the right, then to the left, so that all sides might be heard. The " palaver " had commenced about nine o'clock, and at mid-day sun only four speakers had been heard. The fifth, who was an orator of some repute, rose from his stool where he had been reclining, drank from the calabash of water handed him by his wife, and then adjusting his loin cloth and picking up his notes—a bundle of twigs as remembrancers of the various points—he stepped forward. With an air of complete mastery of his facts, he sped on quietly for the first quarter of an hour ; at the close of every period he turned to his supporters for approving applause, which was given in a chorus of assenting " Oh's." From calm and reasoned recital of facts, he then passed on to his deductions, and for another quarter of an hour he drove his points home amid the now increasing interest and applause of his own side and the derisive laughter of the opposition.

At the end of half an hour, excitement was beginning to run high. The orator now threw himself into a final effort ; gathering up his facts and deductions, he charged the other side with every species of deception and fraud, and as he did so he danced to and fro with his body bathed in perspiration. Every sentence now was punctuated by the almost frenzied applause of his supporters. In his concluding sentences he made a fervid appeal for justice, all the while moving backward towards his expectant friends and wives. He uttered his concluding sentence with arms waving aloft and then swooned into the arms of half a dozen wives who emptied their calabashes over that quivering perspiring body. This man had never read the trial of Warren Hastings, but I could not help recalling Sheridan as the African orator lay there apparently in a dead swoon—I knew of course that he was inwardly rejoicing in his great feat and in the applause which awoke the echo and re-echo in the great forests immediately behind us.

If this " full dress " palaver fails to secure an amicable settlement, the tribes in the Congo basin do not abandon their efforts. They surround the villages with sentinels and adopt various defensive measures, but before hostilities actually begin, they select a sort of " daysman," who, to act in this capacity, must be of peculiar relationship to both tribes ; that is to say he must be able to claim parentage in both dissentient communities.

The daysman goes forth wearing a fringed and partially dried plantain leaf sash thrown over the shoulder so that the sentinels of both tribes immediately recognize him and his sacred office. It is very seldom this arbitrator fails to secure a peaceful termination of the dispute. If he does fail and hostilities break out causing loss of life, he immediately renews his efforts ; indeed he never ceases that constant passing to and fro on his errand of peace and goodwill.

The proposal to sheathe the sword, or, more accurately, to unstring the bows and cleanse the poisoned arrow heads, is followed by another palaver. It was once my good fortune to be invited to act as arbitrator at one of these interesting proceedings.

The drums in all the surrounding country were beaten at cockcrow and immediately the two tribes, under their respective chiefs and headmen, began marching towards the rendezvous—a clearing in the forest outside the village at which we were staying.

I was rather alarmed at the fact that though this was a peace conference, every member of that great concourse carried not only spears, but bows and arrows, and I knew that the slightest indiscretion would precipitate a bloody fight.

All the old history was retailed again through that long and burning hot day. Once or twice a speaker raised the devil in his opponents ; spears were gripped and arrows snatched from their quivers, but at last better counsels prevailed and terms were agreed upon. The question at issue was a boundary dispute, but lives had been lost and prisoners taken on both sides. The boundary was readjusted to the apparent satisfaction of both parties, prisoners exchanged and compensation paid for the killed on either side—this latter surely an advance on " civilized " terms of peace by the way !

The ceremony of " signing the peace " is not the least interesting part. First a strip of leopard skin was secured and then a bunch of palm nuts. The skin was pinned to the ground by a dagger, and each chief and headman followed me in driving the dagger deeper into the earth. When it was firmly fixed the leopard skin was drawn first one way, then the other, until it had been completely severed. A half was given to a young chieftain of each tribe, and they were instructed to " haste to the river, young men, throw the separated skins upon the waters that all men may know the quarrel is now cut in pieces (*i.e.*, is destroyed)." This done, the bunch of palm nuts was taken and a spear from each party driven into the head of nuts. Two more men were selected, again from each tribe, and instructed to " Carry that head of nuts carefully, young men, throw them into the river that all men may know that our spear heads are buried, that fighting is over and peace made for ever and for ever."

In this exceptional case the " for ever and for ever " only lasted three months ! but in the great majority of such cases peace though threatened is maintained for many a year.

V

THE NATIVE AS A MONEY MAKER

If the African woman is a prudent banker, the man is the money maker. The range of remuneration they receive for their labour is no less divergent than one finds in Europe. The Sierra Leone native will obligingly row you ashore to Freetown in fifteen minutes " for two bob, Sah " ; but his brother paddler on the Chiloango, or the Congo, will paddle for you throughout a week for 5d. a day, coupled with a plump bat or the leg of a monkey by way of rations.

There is one form of money making which is fastening its fell grip ever more firmly upon the middle-class African —money lending. It is extremely difficult to deal with this question in West Africa by legislation, but a good deal can be accomplished in various directions by a watchful administration. One case brought to my notice was that of a cook who was compelled to pay £2 10s. interest on a loan of £4 for six months. Another one was that of a teacher who required a loan of £6, for which he had to pay 12s. per month interest. I was also assured that frequently 10s. a month interest is exacted for small loans of £1. In some parts of the Gold Coast borrowers find themselves in such straits that they are often compelled to pawn their children.

The wages of agricultural labourers vary very considerably. In Southern Nigeria labourers working for native employers receive from 15s. to 20s. per month. The contracted labourers on the islands of the Gulf of Guinea—that is Fernando Po, San Thomé and Principe— are all " contracted " at paper wages, varying from 10s. to 15s. per month, but neither under the Spanish or Portuguese Administrations do they receive more than half their pay when it is due, the other half being placed in the hands of the Curador. In German Cameroons the wage is seldom more than 10s. a month, and more often the labourers only receive 8s. In the hinterland of Belgian and French Congo, the unskilled labourer receives from 6s. to 8s. per month. All these wages are exclusive of board and lodging, but generally a certain amount of clothing is supplied freely. In many parts of the various colonies, however, stores are opened by the plantation owners to tempt the labourer into purchasing goods which usually carry a respectable profit.

The hardest work and the poorest pay falls to the carrier ; that patient burden bearer rarely gets, in any part of Africa, more than about 9d. per day for his heavy task. The Upper Congo was thrown open to the advance forces of civilization by a continuous stream of carriers, who occupied from a fortnight to three weeks reaching Stanley Pool from Matadi, a journey for which they seldom received more than a sovereign a load. " Big money," however, is earned by the cocoa carriers of the Gold Coast, but the conditions are entirely abnormal. The cocoa carrying enterprise as at present organized cannot be other than a temporary expedient and the general army of African carriers will have to be content with a wage varying from 4s. 6d. to 7s. a week.

The African is by nature a trader, and no more honest than many Europeans in his business transactions, and on the whole I am afraid less honest than the reputable business houses of West Africa. It is only fair to say that the native merchants trained under the rigid standard of European firms—particularly the Basel Mission of the Gold Coast—maintain a standard of honest trading which does credit to the firms under which they received their commercial education.

The ambition of most young men on the Upper Congo is focussed upon wives. Without earthly possessions, their only hope of matrimonial bliss is in the death of a relative from whom they may " inherit " a partner, if there is a disparity in age an " exchange " is always possible, subject, of course, to an additional dowry. But this chance is remote and the waiting time is always long, tedious, and full of social complications. One day a young man in the Congo endowed with more than the usual share of courage and trading instinct, hit upon a plan which has for years found increasing favour. The captains of steamers could only with difficulty work their boats up and down that 2000 miles of waterway between Stanley Pool and the great tributaries of the Upper Congo, for lack of wood fuel from the forests. Here, then, was the chance for the enterprising native. He bargained with the white man upon the following basis. To travel with him to Stanley Pool and back again, a journey occupying four weeks, to cut a square yard of wood every night on the journey, and to be allowed to sleep during the day. The wages for this enterprise to be ten francs payable at Stanley Pool, and the free transport back

again of one bag of salt and one box of sundries. This
suggestion, sound in its common-sense, giving the white
man fuel without trouble, was promptly agreed upon,
and with ten others on the same terms the contract was
confirmed. The white man went to his bunk that night,
happy in the thought that for one journey at least he
would be saved the eternal " wooding palaver." The
native youths, too, went to sleep, and possibly dreamed
of the wedded bliss which was now so unexpectedly
within sight.

Four weeks later the " Stern Wheeler " returned and
put the respective wood cutters ashore at their different
villages, each with a bag of salt and a few sundries purchased at Stanley Pool with the 10 francs. The eyes of
certain comely young African women shone brightly that
night as they heard of the brilliant enterprise of their
prospective mates. A few days later two or three
parties in small canoes pushed away from the banks and
started on a ten days' journey up one of the small
tributaries which abound everywhere on the Upper Congo.
In each canoe were precious bags of salt and a tiny spoon
for retailing the " white powder " to distant tribes. A
fortnight later family palavers were held and a sufficient
dowry laid at the feet of the damsel's father. The
nightly wood-chopping enterprise had produced 10
francs which had in turn obtained a bag of salt, a hundred
common safety pins and a cheap mirror. The salt and
pins had disappeared and there lay on the ground in their
place the coveted dowry of £2 sterling in native money
for the father, and a mirror for the mother of the
native bride who now gladly joined her husband for

better or for worse. There is your African trading instinct !

Since that day many a young man has followed that example, but with competition dowries have risen and the value of European produce fallen. Nevertheless, to-day, many a native on the Congo waterways is cutting firewood to and from the ports in the hope of raising the where-withal to obtain his heart's desire.

It is said of the Indian coolie that anywhere he will make two blades grow to the one blade the white man can produce. In this respect the African follows hard on the heels of his Indian rival. The white man will often select what seems a most promising piece of land, but for some reason his crops fail. The native will choose a little out-of-the-way patch and cultivate it in a style which calls forth a pitying, almost contemptuous smile from the white, but somehow that native has struck fertility and his crops flourish amazingly.

In Southern Nigeria I met several successful native farmers, who seem in some respects to outdo their friends in the neighbouring colony of the Gold Coast. One of these had some years ago bought 200 acres of land at 4s. per acre, and soon it was discovered that he had obtained a very fertile patch and he was offered no less than £5 an acre and his crops at valuation, but Mr. X. has a keen business head upon his shoulders and finds it more profit-able to cultivate cocoa, palm nuts and rubber than to sell his land even at an enhanced price. Every time he makes a few pounds he extends his plantation, "pulls down his barns and builds greater." This man has now a turnover of nearly £20,000 a year.

There are scattered all down the coast in British colonies native traders pressing on to positions of dominating influence. These men can handle cargoes of four figures and pay at an hour's notice. They receive regular cable information of the prices of different commodities on the European market, and several of them have branches which connect by telephone. Most of them conduct their business on modern principles with typists, cashiers, messenger boys and so forth. Not a few of them are frequently in a financial position to strike a bargain and settle a transaction before the European firm can get a cable reply from the home directors. They are up-to-date traders in being able to supply anything which may be demanded of them, or if not in stock they will promise it—and keep the promise—on a given day. If an order is specially urgent and has to come from Europe, a messenger will meet the ship, take off the package and deliver it to the client within an hour or two of the ship's arrival. One of the most interesting transactions I know of occurred in a certain British colony. A chief, for some reason, was in great need of a large elephant's tusk, and after fruitless endeavours to obtain one, a native trader relieved the old man's anxiety by offering to deliver a tusk the required size, to cost about £80, within a month. Promptly to time the tusk was delivered— the cute trader had cabled to Europe for it ! " Holts," " Millers," and other all-wise competitors in that town knew how imperative it was that this old chief should have a big tusk, and I was told they tried their " up country " stores, but it never occurred to them to order from Europe. There again is the African trading

A NATIVE PLANTER IN HIS FUNTUMIA PLANTATION.
SOUTHERN NIGERIA.

RUBBER COLLECTORS, KASAI RIVER, UPPER CONGO.

instinct, which put a clear £10 note in the trader's pocket !

The legal profession is beyond question the most lucrative in West Africa, but this does not obtain in Africa alone. The mass of the people have not yet learned to settle their troubles without the aid of the legal community. The fees paid to the coast barristers are surprising. I was informed that in one colony more than one native barrister has an income of close on five figures. I had no reliable evidence upon this and should think it an exaggeration, but the style in which the coast barrister lives and moves must certainly require a substantial income. Certain it is too that none are more generous with their money.

Unlike the medical profession, no colour bar stands between the barrister and the free exercise of his ability. Surely the position of these medical men calls loudly for redress, the profession which, above all others, is needed in the fever-haunted colonies of Africa, yet between the increase of these men and the countless sufferers there is firmly fixed the detestable colour bar of prejudice.

Though the native has not yet become convinced of the safety of banking, the sums placed by them on deposit in the three British colonies—Sierra Leone, the Gold Coast and Nigeria, are nearly £80,000.

When we reflect upon these natives rising to positions of greater power and influence in British colonies, and when we are prone to criticize British administrations, it will not hurt any of us, either native or European, to remember that less than a century ago these centres were amongst the principal slave markets of the world.

VI

THE AFRICAN WOMAN

THERE is assuredly no country whose women are more interesting than those of Central Africa. Certainly there can be no place on the habitable globe where women are so continuously industrious. Amongst African women there are no unemployed and no unemployables. In all the hinterland, the women are the agriculturists. In the early morning, often before sunrise, they file out of the village to their plots, perhaps a mile away from the town, where there is always something to do ; weeding and planting being almost an integral part of the daily routine. When the gardens have received attention, meals must be considered and the woman proceeds to dig up the manioca tubers, but only to bury them beneath the water in some forest stream or pool to extract the injurious element. In a few days hence the load of sodden tubers will be ready for the native culinary art.

Ten minutes in the forest and the woman has gathered the fuel required for her cooking ; then loading her basket with the manioca left to soak six days before, she places a layer of leaves between it and the firewood, and shoulders her burden. She steps out brightly for home, in company with perhaps another twenty matrons.

It is not every day that she is able to finish by noon, for in the planting season the gardens demand her labour for whole days at a stretch. Some weeks before the husband has perhaps started a new field by cutting down at immense labour hundreds of trees, which lie there scattered in all directions till the tropical sun dries up the leaves and smaller branches. Then a torch at one end of the clearing starts the whole area in a blaze.

It is at this stage that the wife comes along with her seeds and cuttings, digging little mounds all over the area and raising the soil by heaping upon it the cinders, dead leaves and ash, which provide the only manure these primitive folk possess. Between the rows of manioca she may plant gourds, Indian corn and ground nuts, and thus secure a general crop all over her cultivated field.

From Sierra Leone right away to the north bank of the Kasai, these domestic crops vary but little, but on arriving at the southern bank of the Kasai, the change becomes very marked, for the extensive fields of manioca and cassava give way to mealies as the staple food.

The field is the first charge, so to speak, upon the time of the African woman, but to her belongs also the major responsibility for providing the daily meals. The primitive African is almost a vegetarian, though he dearly loves meat. Trapping edible fish is by no means frequent, and the wife, knowing with her civilized sister how important it is to feed the man, will often snatch an hour or two from her busy life and run to the nearest stream and catch some " small fry," with which to

make savoury the evening meal of cassava and pottage.
In season she will hunt through the forests for the caterpillars which abound on certain trees and which by
some tribes are regarded as great delicacies, particularly
those tribes inhabiting French and Belgian Congo and
the Cameroons. The Gold Coast people substitute large
snails, of which they appear inordinately fond.

There are four principal dishes which, with slight
variations, prevail throughout Western Africa :—

1. There is the staple food of manioca, which is
sometimes boiled and pounded into puddings, resembling
a lump of glazier's putty. Cassava or sweet manioca is
never soaked, but cooked fresh from the ground and is
much liked by Europeans.

2. The plantain, which is prepared in many forms by
roasting, baking, frying and boiling.

3. There is pottage, the body of which is composed of
pounded leaves from the manioca plant, closely resembling spinach. In most parts of the tropics, green
Indian corn is introduced freely into this dish.

4. There is the palm oil chop, which, as I have shewn
in another part of this book, is not a " chop " at all, but
anything from a caterpillar or a beetle to the leg of a
dog or buffalo.

Perhaps next in importance to the position of agriculturist is that of cook. Give the African woman a
clay pot, a pestle and mortar and a few leaves, and she
will produce in quick time a meal which even a European
can relish. She is a trifle too fond of chili peppers and
palm oil for a sensitive palate and fully believes that a
fair proportion of earth and other etceteras add to the

WOMEN POUNDING OIL PALM NUTS.

flavour and digestibility. Her husband, with a natural weakness for chili peppers and oil, and himself not averse to " foreign bodies " in his food, readily consumes nearly two pounds of prepared manioca and pottage at a single meal.

With cockcrow, the woman rises, steps outside the hut and in lieu of washing herself, yawns two or three times, then stretches herself in several directions, and is ready for the day's work. She will first sweep her hut, open the chicken-house, pluck a few dew-covered leaves to wipe over the faces of the children, and then pick up her basket and set out for the gardens. Returning, she will pull the fire logs together and again shoulder her basket and go off to catch fish, or to hunt caterpillars. Some of the older wives may stay in the village to fashion clay cooking-pots, weave baskets and mats, or crack palm kernels.

About four o'clock, " when the monkeys in the forest begin to chatter," the women return to their huts and commence preparing for the principal meal of the day. Above the hum of conversation, the passing jest, or the humorous repartee, the clear ringing thud, thud, of pestle and mortar is distinctly heard. Most dishes at some stage or the other are pounded. The boiled manioca, the pottage leaves, the palm nuts, the plantains, all find their way to the mortar, and no doubt the muscular physique of many of the women is largely the result of the perpetual wielding of the heavy wooden pestle.

The African woman is at home with any industry, hardly anything comes strange to those deft fingers and muscular arms. The husband may go on a journey by canoe, and his wife, or wives, will be there paddling amidships and cooking the meals at intervals. The husband, however, always takes the post of danger, which may be bow or stern, according to the weather, the current, or the district through which they may be passing.

Much has been written, backed by little knowledge, about the brutality of man in making the woman carry the loads when on an overland journey. To the un-initiated European, it may seem callous for a strong able-bodied man to walk in front of a line of women every one of whom is struggling along with a 50-pound load on her back. But make them change positions, force the man to take the load, tell the women to walk in front, and before you have gone many yards the women will all have bolted into the bush, for the " Lord Protector " under a load is no longer ready to shield them from the danger which lurks behind every tree and beneath almost every leaf in the African forests. The African knows his business when on a journey, and his first duty, from which no matter what the odds, he never shrinks, is that of protecting his family from the ravages of wild animals no less than the violence of hostile tribes. But to do this he must be unencumbered and alert.

The women of West Africa, by reason of their thrifty natures, are frequently the bankers. To them the husband entrusts the keeping of his worldly goods, and right sacredly they guard anything placed in their keeping. Not only are the women trustworthy bankers, but as moneymakers they are extremely keen. The finest business woman it has been my lot to meet was a farmer

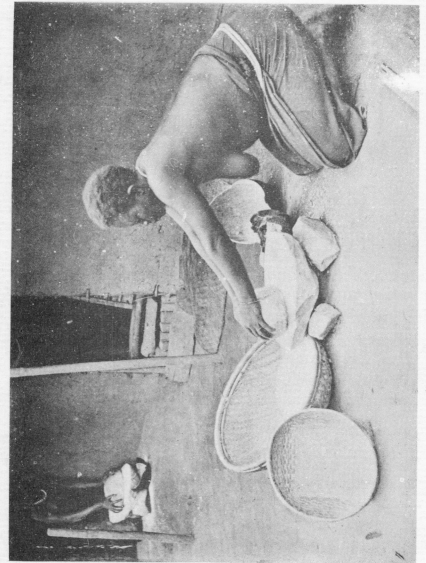

GRINDING CORN ON THE KASAI, UPPER CONGO.

woman of Abeokuta. This old lady could tell at sight, almost to a penny, the value of a pile of kernels without weighing them. I fell to discussing with her so technical a question as the possibility of cotton in Southern Nigeria, and she was adamant in her opinion upon this : " Unless they can guarantee me 1d. per pound for unginned cotton and 4½d. per pound for the ginned, I would even prefer to grow yams." I gathered that on the whole she was not likely to become a shareholder in any cotton producing company.

In the mart the women excel. It may be in the streets of Accra, Abeokuta, Freetown, or in that finest of all marts in West Africa—Loanda, or again in some wayside market of a tributary river in the far distant hinterland. Wherever you find the market, the women are in control and right merrily goes the auction. The din amounts to a pandemonium, the tricks of the trade are to be looked for in every basket of fruit or pile of vegetables. The eggs are probably old ones, carefully washed and possibly doctored ; that fowl tied by the legs could not walk from sickness if it were free. Billingsgate, Smithfield, Covent Garden, rolled into one could not be at once more entertaining, more noisy and more novel than those African markets where you may buy almost everything you want, and receive a great deal gratis that is not welcome.

THE AFRICAN WIFE

Is there any feature, social, political or religious so important in West Africa as the wife and mother ? No " teeming millions " are to be found in the African

tropics and every colony is crying out for more native
workers as the development of her industries gets beyond
the fringe. As a wife the African woman is generally
but one of a number. In most coast towns to-day the
stress of modern competition has forced up the cost of
living, which together with the absorption of civilized
ideas has made monogamy—oftentimes, alas, only surface
monogamy—the passport into respectable society. But
away from the coast towns, though it be only a few miles
away, polygamy is prevalent almost throughout West
Africa.

Christian converts profess an abhorrence, and in
many cases I am satisfied a sincere abhorrence, of polygamy, but the fact remains that this causes more trouble
in the Christian Churches in West Africa than all other
evils put together. In the purely pagan areas there is
no doubt that the woman regards polygamy as a desirable
condition ; she argues that the position of the husband
is gauged by his many possessions—wives and cattle, and
that she prefers being the wife of a great man to that of
some insignificant fellow who can afford to keep but
one ! Again she will point out, and with obvious truth,
that if a man possesses several wives, the burden of
agriculture, of fishing, of kernel-cracking, and the domestic
duties spread over four, five or more persons is proportionately lighter upon each individual.

Into the sentiment of polygamy there is also the
practical consideration of offspring. No matter how
plain the daughters, no matter how slightly cicatrized
they may be, no matter what imperfections the boys
may have, if they are the children of a much married man

A CHRISTIAN COUPLE RETURNING FROM THE
GARDENS TOWARDS SUNSET.

WEAVING CLOTH IN THE KASAI, UPPER CONGO.

they are certain to make "good matches." The sons may be certain of securing the daughters of chiefs no less famous than their father ; if a girl, her dowry will not be her intrinsic worth, but will be gauged likewise by the position and possessions of her father.

It does not seem to be generally recognized that there is both voluntary polygamy and in a very real sense obligatory polygamy. A man inherits wives from his father or uncle, just as he inherits other possessions. In most cases of course he gladly accepts his inheritance. This, I know, is a revolting custom to the European, but to the African not merely desirable but the only honourable future for his father's wives. His own mother reigns as a sort of dowager Queen in the household and keeps order in the harem of her son. I have often discussed this feature with the women themselves and find that invariably they regard any other course with the utmost repugnance. Why, they say, should they suffer the disgrace of being passed on to other husbands ; what evil have they done that their rightful husband should disown them and refuse to accept them as his wives ? his father loved and cherished them, and why should the son disgrace his father's name by refusing to follow in his steps !

In one or two cases Christian men have actually put away wives whom they have inherited in this manner, but the women concerned have always felt that the shadow of disgrace has fallen upon them and that they are outcasts from the social life of the tribe.

This custom, like most extreme polygamous concomitants, finds its fullest development in the upper reaches of the Congo river, but it is also found practically throughout the whole of the Congo basin.

The general attitude adopted by missionaries in West Africa is that of rigidly excluding the husband of more than one wife from Church membership, and this no doubt accounts for the apparent lack of success which statistics seem at first sight to demonstrate. Almost every missionary, however, will point out to the traveller, man after man who, though not a member of his church is, he declares, with a regretful sigh, "more of a Christian than the majority of our members." The German Basel Mission in the Cameroons excludes all polygamists from Church membership and they have been fortunate in obtaining King Bell as a monogamist member. In the " oil rivers " of the Niger, the same rigorous position is taken up by the missionaries.

In not a few churches in Southern Nigeria, polygamists are certainly admitted to membership of the churches. These men if not openly polygamous are notoriously so in private life.

The Christian Church has, in polygamy, a problem which at present defies solution ; the custom is so much an integral part of African life that a conversion to Christianity involves an abrupt termination of the convert's former habits, the effects of which reach far beyond the individual most intimately concerned. One of the greatest difficulties is that of the outcast wives. In one Mission in Southern Nigeria if a man becomes a Christian convert he is asked to call his wives together and explain his position, then to select one, put the others away and provide for their maintenance. But even this involves a sense of injustice and is, I am told, fruitful in many cases of deplorable results. The women thus set aside regard themselves not unnaturally as outcasts, as they have lost the affection of their husbands and are therefore in disgrace. In many cases, I am told, these women become either temporarily or permanently the mistresses of other men who do not hesitate to taunt them with the fact that they are outcasts from ordinary native society.

No doubt there are exceptional cases where women so put away find mates amongst the bachelor members of the Christian community, but even these young fellows—and more particularly their parents—are not always over anxious to accept as a wife for their son the woman whom another man has set aside.

The Honourable Sapara Williams, one of the ablest men in West Africa, expressed the opinion that it i. imperative the Christian Church should find some other solution than exists to-day for this difficulty if it is to maintain and increase its hold upon the native tribes of tropical Africa. We see already a native Christian Community in Southern Nigeria known as the African Church existing avowedly upon a polygamous basis and growing rapidly in membership and influence. This Church is entirely self-supporting and is becoming more and more propagandist. In the course of time it may easily produce what will be called an " African Wesley," or an " African Spurgeon," and the result we can foresee. The African *en masse* is inflammable material and intensely patriotic ; let such a man emerge from their ranks and the doctrines he preaches will spread like wildfire.

It is universally recognized that in case of any modification of the attitude now adopted by the European government of Christian Churches, thousands of adherents would be secured in every colony. The heroic attitude hitherto adopted surrenders to Mohammedanism a potent factor in the propagation of its beliefs, hence the extraordinary advance made by the apostles of the prophet.

There is evidence that the position maintained by the Christian Churches as a whole upon this aspect of its work leads to widespread immorality amongst Church members, but wherever it becomes too notorious, the delinquents are, with certain exceptions, excluded from membership. It will be readily seen therefore that should any single Christian denomination once lower its standard in this respect, converts would flock to it in thousan ds. The African Church does this, and springing from the people themselves, meets the situation. Its members probably represent the Christian natives of the near future in Southern Nigeria, men for the most part commercially successful, boldly solving their own problems, living an easy-going and comfortable life, their religious standard lowered to their own desires. Can we criticize them ? If we do, we must beware, for they will tell us that it is more honest to live open polygamous lives than the fraudulent lives of professing Christians— white and black—whose hypocritical attitude, particularly on sex questions, is a by-word on the West Coast of Africa. I fear there is too much truth in this retort. White men, at least, must hold their peace, and there lies the greatest danger !

It is generally accepted that polygamy is productive

of a high birth rate, and Sir William Muir has given this as one reason for the almost miraculous advance of Mohammedanism. It may have been, and may still be true to-day of Mohammedanism that polygamy produces a high birth rate, but that existing polygamists in tropical Africa to-day produce a greater number of births than monogamists is, I am satisfied, open to serious question. At the same time I think it is clear that prior to European occupation, polygamist Africa maintained a higher birth rate than is possible under modern conditions.

The reason for this is not far to seek, for the chiefs, possessing as they did unrestricted power over the community, could terrorize into complete submission every unit of the tribe. Wherever polygamy existed the wife was kept faithful to the one husband by the knowledge that unchastity was forthwith rewarded by instant death. The young men also knew that a *liaison* meant either that they were sold into slavery, involving in all probability ultimate sacrifice, or they would be hanged on the nearest tree.

This is so even to-day amongst those tribes beyond the reach of white men.   One day, when crossing towards the main Congo river, I suddenly heard wild shrieks from a person evidently in great danger. Rushing to the spot, I found a woman bound hand and foot, and standing over her was a burly young chief with an executioner's knife raised aloft. In a moment more that woman's head would have been hacked off had I not promptly gripped the man's arm. With a terrible oath he attempted to spring upon me, but the head-men of the village, who had also hurried to the scene, fell upon him and wrenched the knife from his hand. For a quarter of an hour nothing would stay the man's fury ; it took six of us to hold him. Ultimately, however, he calmed down and explained to me that his wife had been unfaithful and that she merited the death penalty. I gave him some presents to appease him further and he agreed to forgive the woman if I would " reward him." As the gift he asked was to me a trivial matter, and the only chance of saving the woman's life, I gave it to him. The woman herself, in gratitude, at once wrenched from off her wrists a bracelet which she presented to me as a keepsake. I fear, however, that after I left the village, she suffered a cruel death for her unfaithfulness. It will be readily seen that these conditions are only possible in regions where there is no restraining hand.

The question of the birth rate under monogamist and polygamist marriages in West Africa has always been of absorbing interest to me and my diaries are full of jottings bearing upon the subject, but very few are worth a permanent record. Amongst the Christians of Accra many monogamists have considerable families and from personal observation twins appeared to be fairly frequent. In the hinterland, we were informed, that the " baku "—or tenth child—is by no means rare amongst the " Twi " people. The largest family we found amongst the monogamists of the Bangalla region of the Congo was five children, the average appearing to be three. But West Africa is very weak in reliable statistics.

In our recent journeys, I selected four areas and obtained with some accuracy the composition of several groups of villages. It was impossible to accept the figures from some districts because the people, fearing there was some subtle move behind our requests, either gave evasive replies or figures which were obviously inaccurate.

The following six groups, however, are reliable. They were gathered from areas hundreds, and in one case over a thousand miles apart :—

Five Hinterland Villages of the Kasai.

	Men.	Women.	Average woman per man.	Offspring. Boys.	Girls.	Total.	Average per man.	Average per woman.
A . .	241	316	1·315	70	81	151	0·626	0·477
	(201 monogamists)							

River-side Villages on the Upper Congo.

	Men.	Women.	Average woman per man.	Boys.	Girls.	Total.	Average per man.	Average per woman.
B . .	26	45	1·875	14	15	29	1·115	0·644
C . .	41	54	1·317	5	4	9	0·222	0·166
D . .	16	42	2·625	10	6	16	1	0·380

Hinterland Village, Upper Congo.

	Men.	Women.	Average woman per man.	Boys.	Girls.	Total.	Average per man.	Average per woman.
E . .	31	69	2·225	23	20	43	1·387	0·623

Remote Hinterland Village, Upper Congo.

	Men.	Women.	Average woman per man.	Boys.	Girls.	Total.	Average per man.	Average per woman.
F . .	196	319	1·627	171	148	319	1·627	1

In group " C," the principal polygamist possessed fifteen wives, but only two children. Sixteen monogamists had no children.

Group " A " is taken from the Kasai, where monogamy most widely prevails, but of the two hundred and one monogamists, one hundred and three had no children. The principal polygamists possessed six, eight and thirteen wives respectively. The two first had no children at all and the chief with thirteen wives had two boys and three girls.

From these figures no deduction is possible as to the advantage of either polygamy or monogamy upon the question of birth rate. One deduction only is clear.

The birth rates in the following order are with estimated distance from effective civilized Government :—

	Average birth-rate per woman.	Distance from effective civilized Government.
Group C . . .	0·166	10 minutes' walk
,, D . . .	0·380	30 ,, ,,
,, A . . .	0·477	1 hour's ,,
,, E . . .	0·623	1 ,, ,,
,, B . . .	0·644	1½ ,, ,,
,, F . . .	1	2 days' ,,

The birth rate figures are lamentably low, and being selected from areas so widely apart give anything but an encouraging indication for the future of the Congo. The deductions from these figures is unmistakable and only confirms what one hears everywhere, not only in the Congo but all over the West Coast of the utter demoralization which is flooding these territories.

The Congo is by far the worst. Europe was staggered at the Leopoldian atrocities and they were terrible indeed, but what we, who were behind the scenes, felt most keenly was the fact that the real catastrophe in the Congo was desolation and murder in the larger sense. The invasion of family life, the ruthless destruction of every social barrier, the shattering of every tribal law, the introduction of criminal practices which struck the chiefs of the people dumb with horror—in a word, a veritable avalanche of filth and immorality overwhelmed the Congo tribes.

To-day one sees the havoc which King Leopold created when he let loose upon the Congo tribes the scum of Europe. None have escaped the infection ; girls of tender years and even boys not yet in their teens delight in practices of which in the old days the chiefs would have kept them in complete ignorance for another five years. Upon the women the results have been by far the most revolting, for in the Congo the majority of women have lost their womanhood and have fallen into a daily condition from which even the beasts of the forest refrain.

The truth is that in the greater part of West Africa neither monogamy nor polygamy is the prevailing relationship between man and woman. Doctors, administrators and missionaries all know it, and are all powerless at present to bring the situation under control. It is useless for the administration to make laws for practices beneath the surface, the only thing the officials can do, and should do without delay, is to see to it that an ever higher example is set to the natives. This is where the Belgian and French Congo officials have failed so utterly.

The Christian missionary alone touches the evil, and though he is defeated again and again, he plods steadily on preaching a perfect chastity—too lofty a standard for most natives at present—but without doubt gathering round him an ever increasing number not only of men but of women who, apart from occasional lapses, set a bright example to the whole countryside.

The birth of children is in primitive Africa rarely attended by anything abnormal. If a native nurse is confronted with complications, she immediately throws up the case in despair and appeals to the witch doctor, but normally the birth of children is taken as quite an ordinary part of the daily life. One day we were passing through a native village, and there, lying on a plantain leaf, were two chubby little twin girls but half-an-hour old; the mother was sitting close by "resting." This picture was so beautifully simple that my wife went with a boy to bring up the camera and plates, but on arriving at the spot in about twenty minutes the woman had picked up her twins and carried them home! That is primitive Africa, but in the coast towns where African womanhood delights in corsets and other European follies, the suffering at childbirth is in many cases almost as acute as that amongst the European community. With several Congo tribes, the belief is firmly rooted and put into practice that in order to change the colostrum flow to that of milk, co-habitation is essential.

With many tribes throughout West Africa, the period of lactation is prolonged; frequently the mother nurses the child until it is two, three, and even four, years old. A case of adultery was brought before the District Commissioner's Court at Ikorodu in Southern Nigeria in April last year, and in the evidence it came out that the accused woman was suckling a child four years of age. The District Commissioner ordered her to cease nursing the child within three months.

The death rate amongst the young children in West Africa is very high and no doubt arises from the deplorable manner in which they are brought up. There is practically no attention given to diet or cleanliness, with the result that any disease which attacks a family quickly spreads through the community.

Amongst the Dagomba of the Northern territories of the Gold Coast colony, the woman who has given birth to a child leaves her husband's compound and goes to that of the father-in-law, taking the child with her, where they stay for a year. At the end of this period the wife and child return to the home of the husband and father.

TWINS

It is a mistake to assume, as some writers do, that the taboo on twins is a prevailing custom amongst West African tribes. The distribution of the taboo is extremely erratic. Twins are unwelcome in the Northern territories of the Gold Coast, yet the reverse is the case amongst the Egbas of Nigeria. In the Congo territories, twins cause the greatest joy to a tribe and the mother is lauded wherever she goes, whilst amongst the tribes of the oil rivers of Nigeria, the birth of twins is regarded as the most fearful calamity which can fall upon the community.

In the Upper Congo regions, the traveller may frequently see two earthenware pots hoisted on forked stakes which have been driven in the ground, one on either side of the path, and these are in honour of twins born in the nearest compound. Every person passing by those pots will religiously pluck two leaves and throw one at the foot of each forked pole as a votive offering to " Bokecu " and " Mboyo," as all good twins are named.

The tragedy of the oil rivers is one of the most distressing in West Africa. Throughout the Eastern, and to a considerable extent of the Central Province, the cruel custom prevails of putting to death one, sometimes both twins. The British Government spares no pains in the effort to combat and overcome these practices, but though much good has resulted, the custom still holds its own.

Not only are the children killed, but the mother is immediately driven from home for she is no longer regarded as a chaste woman and rapidly becomes an outcast from Society, living upon the proceeds of prostitution. In some districts, however, this custom is less rigorous, and the mothers of twins are allowed to form isolated villages and to engage in trade. Some tribes, again, whilst driving them from the homes of their husbands, permit them to engage in agricultural pursuits upon the husband's lands.

The missionaries are doing much towards weaning the tribes from this murderous practice. One missionary society working amongst the Ibunos, a tribe of five thousand people, claims that through the conversion to Christianity of a large section of this tribe, the horrible practice of murdering the twins and making the women outcasts has ceased. It is of course difficult to control absolutely a statement of that kind, but it is only the Christian missionary who can hope to deal effectively and permanently with a subterranean evil like twin murder.

An interesting custom which survives in the Upper Congo is that a man may never see or speak to his mother-

"TWIN POTS" HOISTED ON FORKED STICKS EITHER SIDE OF PATHWAY,
IN HONOUR OF NEWLY BORN TWINS, BANGALLA. CONGO.

in-law, and should he by accident turn a corner in the village compound and meet her face to face, he must at once send a propitiatory offering. If she should come into the house where he is sitting, he will promptly raise a mat and hold it between them, so that they may not see each other.

On the whole the lot of the African woman is a hard one. She has her occasional pleasures it is true, but from childhood hers is a lifelong drudgery with, however, the one sure recompense, that in old age it is the joy and the privilege of the younger generation to support her.

PART II

CIVILIZATION AND THE AFRICAN

I

THE WHITE MAN'S BURDEN

THERE is a type of African traveller who, hurrying to the coast and back again, returns with all the assurance of a long experienced person to pontifically declare that the unhealthiness of West Africa is all moonshine, that if a man dies it is due to his excesses rather than to the climate. There is, of course, a grain of truth in this assertion ; cocktails, midnight oil and habits of a worse type, undermine the constitution in a manner which leave little resistance to the climatic diseases. Yet after all, tropical Africa is a death-trap.

Some of these assertive and incredulous persons have themselves been badly punished for their advertised temerity. The story goes of one lady who, after having published much nonsense on this subject, was bundled off home in an ice pack ! I know one man who, after a year or two of good health, gave rein to his opinion in the columns of the *Times ;* this good man was no believer in short and effective service, followed by a well-earned period of leave ; he advocated long terms of residence as the certain road to immunity ; that man spent a single term in Africa, towards the close of which the climate made such inroads upon his constitution, that he was never allowed to return.

Those who feel inclined to trifle with and ridicule the dangers attendant upon life in Africa should spend a solid year in some lonely post directing a staff not always amenable to discipline; should live in that comfortless bungalow; should endeavour to tempt the appetite day after day with something from a tin which, no matter what it is called, invariably has the same taste. Then probably a fever intervenes and the lonely resident goes to bed with limbs racked with pain and a head throbbing like the puffing of an express train. By this time the supercilious writer would be brought to know that after all, the climate of West Africa is not that of the Swiss lakes or the Austrian Tyrol.

It may be a melancholy undertaking, but all whites going to West Africa should brace themselves to the duty of visiting the cemeteries. What a story the graveyards of West Africa tell! The fair young lives laid down for the comfort of posterity. Men of all walks in life are there—the official and the trader, pitiably aloof in daily life, now lying side by side; they are there from every profession and trade, the engineer and the miner, the planter and the doctor, the young wife and perhaps the new-born infant. Africa—always cruel —has taken them in the very flower of their manhood and womanhood.

On the Gold Coast I one day walked into the cemetery and standing in one spot recorded the ages inscribed on twenty-seven of the surrounding tombstones; the oldest amongst the deceased was only forty-six years, and amongst the youngest, two had succumbed at the early age of twenty-two. The average was exactly thirty-two years. Not a few had inscribed upon the tombs such information as, " After two days' illness." " After only three weeks in the colony." " After three days' illness." " Died on the way to the coast," and so forth.

One interesting feature about this cemetery is that it is enclosed with a stone wall, about four feet high, and all white men may be buried within the compound, as also respectable natives—respectability, so my native guide informed me, being determined by church-going. Natives, therefore, who were not attendants at church, were buried " outside the wall." Looking over I could see some scores of graves of natives who, not having attended church in life, were divided in death from the church-goers by a foot of stone wall.

Merchants and missionaries would do well to watch more closely the mortality returns of Government publications, for there alone may be seen recorded the effect of furloughs on the health of Europeans. In the slow moving times of twenty years ago, men went to the coast for long periods, and many a missionary and merchant stayed until he died. Government officials, too, were kept at their posts until death carried them off, or they were invalided beyond the possibility of a return. It is instructive to note that mortality is much lower among Government officials, arising beyond question from the fact that they serve short periods, generally of one year only, and then take a furlough in Europe. For many reasons the figures for the years 1901 and 1910 may be regarded as average records. The death-rates among the whites in the two colonies of the Gold Coast and Southern Nigeria, showing a remarkable improvement, are as follows :—

	Southern Nigeria. Death Rate.	Gold Coast. Death Rate.
1901.—Officials	24 per 1000	34·96 per 1000
Non Officials	47·1 ,, ,,	56·30 ,, ,,
1910.—Officials	6 ,, ,,	11·41 ,, ,,
Non-Officials	(not available)	16·52 ,, ,,

Most merchants argue that they cannot afford to bring their men to Europe for short furloughs every year, but one or two good houses are making the experiment with not a little satisfaction to themselves in more than one direction. In the first place a better type of man offers for a short agreement, and then there is the consideration that by preserving the lives of those they have trained, merchants thus avoid the constant re-equipment of new men, the cost of which is very considerable. Nor is the financial aspect the only feature which is proving satisfactory. These merchants find that they reap great commercial advantages over their competitors by being able to hold more frequent consultations with their men. After all, the incidence of cost in connection with passages to and fro is comparatively insignificant on the whole expenditure of the far-reaching commercial enterprises of West Africa.

To preserve the white man's life in Africa, other elements are equally essential. The dwelling-house, recreation and provisions are features sadly neglected by the majority of whites.

There is so much monotony, so much to irritate and to depress in West Africa, that everything Governments

WILD FLOWERS GROWING ON TRUNK OF FOREST TREE.

"THE STORY THE GRAVEYARDS TELL."

and merchants can do to brighten the lives of their employés should be done. The prettiest and happiest of homes are without doubt in German and Portuguese colonies. In both cases it is due, to a very large extent, to the fact that these nations give every encouragement to the taking out of white women, whose very presence, flitting to and fro in the essentially light garments of the tropics, give more than a touch of poetry to surroundings already anything but prosaic.

The Portuguese love of a garden adds to the attraction of their homes ; grape vines are tastefully grown where the Englishman would throw sardine tins ; there is a fernery in one corner of the garden, a rose bower in another, luscious fruits and tempting vegetables grow everywhere in exquisite profusion.

The Germans in Cameroons set aside a colonial fund called the " Widows and Orphans Fund," and I am told it is from this capital account that men draw subsidies with which to take their wives to West Africa !

One of the prettiest incidents I ever saw in West Africa was at Victoria in the German Cameroons. The planter came galloping home from the plantation, and giving a whistle to announce his return, a daintily dressed little matron skipped out lightly to meet him, and arm in arm they walked into a charming little bungalow gay with fern and flower. A few minutes later I passed by the open door and caught a vision of a snowy table cloth, bright with polished silver and glass. I could not help contrasting this with the British factories with their more or less dilapidated dwelling-houses, most of them very dirty, and the general atmosphere in keeping with the slatternly black woman leaning against the cook-house door.

Recreation in some more healthy form than cocktails and billiards is of no less importance than the well-ordered house. In many colonies now there are golf and cricket clubs, but these are only possible in the more civilized towns where there is a considerable congregation of whites. The man who suffers most from fever and despondency is the one stationed at some isolated post of the hinterland. Happy, indeed, is the man with a knowledge of, and love for, a garden ; it will keep his mind calm, provide him with healthy exercise, and a supply of fruits and vegetables which will keep him in good form for his daily routine.

Given a good home, sound mental and physical recreation, short periods of service with proportionately shortened furloughs to Europe, the white man's burden in Africa, to which so many succumb to-day, would be materially lightened, and both white men and women could go forth with a fearlessness which, tempered with care, would largely remove from West Africa the stigma of " the white man's grave."

LIGHTENING THE WHITE MAN'S BURDEN

THANKS to Mr. Chamberlain, a great stimulus was given to the work of rendering the burden of West Africa somewhat lighter. At his inspiration men began to study more seriously the question of dwelling houses, the use of medicines, and the supply of fresh food.

Sir Alfred Jones, Messrs. John Holt and Messrs. Burroughs & Wellcome, have each in their respective spheres spent large sums of money experimenting in various directions, in the hope that science applied to the practical side of daily life and travel would ameliorate, if it did not remove, the distressing effects of malaria.

The trader of twenty years ago lived—but more frequently died—in a wattle and daub house. These I know from experience can be made comfortable, but more often than not they are so damp and insanitary that fever may be looked for every few months. Inside two and a half years, I experienced no less than seventeen fevers, the majority of which were I am convinced entirely due to the wretched habitation in which we lived.

To-day few men live on ground floors, for the mud or bamboo house has given place to the airy bungalow fashioned on brick piles, permitting a current of air to pass beneath which keeps the house dry and sanitary.

It also has the not inconsiderable advantage that snakes and other reptiles which abound in the tropics do not so readily find a lodging as in the mud and sun-dried brick houses of the earlier days. Another improvement which is yearly growing in favour is that of gauze doors and windows which give some protection from the torment of mosquitos and tsetse flies.

On the island of Principe, the doors and windows of almost every house are fitted with gauze, the object of which is to prevent the spread of sleeping sickness which has of recent years overwhelmed that island. The germ-impregnated fly is nowhere in Africa so numerous and vicious as upon that wretched Portuguese island, where few of a ship's passengers care to land, for the risk of becoming inoculated with sleeping sickness is a very real one. Whilst on that island we had to keep an extremely vigilant watch upon the terrible tsetse flies which gave us no peace, so anxious were they to taste our blood. The fly, which is found in most parts of West Africa, is most prevalent in the Bangalla region of the Congo and on Principe Island. In the latter place they literally swarm. There is no buzz to warn of their approach, and usually the first intimation the traveller has of their presence is the sharp stab, followed by acute irritation and swelling. In spite of the precautions taken on Principe, there seems very little hope that the population can be saved from this terrible scourge. In one month (June, 1910), out of a population of 4000 souls, no less than fifty-six perished from sleeping sickness; that is at the rate of 168 per 1000 per annum. No wonder the Portuguese population is leaving the doomed island.

An experiment which is being watched with keen interest is that recently made by Messrs. John Holt & Company. The directors of this enterprising firm have recently placed two insect proof ships on the West African sea and river journeys. The first of these, the " Jonathan Holt," was launched in July, 1910. This vessel was constructed largely under the advice of the Liverpool School of Tropical Medicine, and the object was that of rendering the passengers and crew immune from the germ carrying mosquito. The " Jonathan Holt," the first of the type ever built, is about 2500 tons and with a dead weight capacity of 2350 tons. She draws only 17 feet 6 inches of water, which permits navigation on the river Niger and also allows her to reach Dualla, the capital of German Cameroons.

The doorways, portholes, windows, skylights, ventilators and passages are all protected with mosquito gauze frames easily adjustable. Double awnings are provided and everything which human forethought can do to render the ship proof against the mosquito has been done.

To Messrs Burroughs, Wellcome and Co., every African traveller owes a debt of gratitude. The excellence and portability of their tabloid preparations have gone a long way to minimize the dangers of tropical adventure. During our travels of over 5000 miles we carried with us a medical outfit which left nothing wanting, either for ourselves or for our paddlers and carriers. For fever, for cuts or bruises, or other inevitable ailments of the tropics everything was at hand ; nothing was lacking for the whole caravan and yet the

total outfit weighed less than twenty pounds ! How
great a difference a Burroughs, Wellcome portable outfit
would have made to Livingstone's hard life.   We carried
another case of " tabloid " photographic materials, and
with these developed nearly a thousand plates.   The
whole outfit, both medical and photographic, was easily
carried by one boy.

It may seem strange to the European that African
travellers and writers lay so much stress upon the question
of food supply.   West Africa for years exacted a terrible
toll from her white residents, which might have been to
a great extent minimized had they been able to provide
themselves with palatable fare.   The late Sir Alfred Jones
determined to do something to make the life of the African
merchants and officials more comfortable in this respect.
He fitted out a few ships with refrigerators and began
in a small way to send some of our staple articles of diet
to the leading ports of the coast.   Men and women too,
sick almost unto death, unable to eat the coarse bread,
the tasteless fish, or the tinned mixtures, were then
cheered and in numberless cases restored by the timely
arrival of an Elder Dempster boat with sterilized fresh
milk, eggs, chicken and mutton.

Only too well do we remember those days, fifteen
years ago, when once on board the ship at Liverpool,
the travellers said good-bye to European diet.   How
different the case now !   Directly the ship casts anchor,
coloured messenger boys, and more often the white men,
come on with orders for beef and mutton, eggs and milk,
chicken and sausages, even game and fruit.   It is a great
day when Elder Dempster's boats steam into port,

hurried invitations go out for dinner and luncheon parties, and once a month at least the pale-faced commercial agent or the anæmic government official is able to enjoy a meal or two which puts new life into his tired body.

Over and above the provisions for the passengers on the steamer, each ship will now carry for sale—beef, lamb, mutton and kidneys ; pheasants and other game ; eggs, sausages, fresh butter and sterilized milk ; potatoes, carrots and onions ; kippers, bloaters and salmon ; grapes, pears and apples—a veritable combination of shops, butcher, dairy, greengrocer, fishmonger and fruiterer !

Usually each ship will carry for sale from 1000 to 2000 lbs. of beef, a couple of thousand eggs, three or four hundred pounds of butter, five hundred blocks of ice and three hundred pints of milk. Festive seasons, too, are not forgotten and Christmas boats carry a large stock of turkeys and geese.

Think for a moment what a blessing the monthly visit of a ship like this is to such foodless places as Boma and Matadi in the Congo, the island of Fernando Po, the isolated merchant houses of Rio del Rey, or the ports of Spanish Guinea ;—the drawn and sickly faces of the men who come off for provisions tell their own tale. They can not only buy all they want, but at a reasonable price. The Belgian in the Congo buys beef cheaper than he can in Antwerp, *i.e.*, 10*d*. a pound. Lamb and steak he can get at 1*s*. per pound. The Scotch engineer running his steamer up and down the Ogowé can get a whole box of Aberdeen haddies for 5*s*., or salmon at 2*s*.

a pound.   Ice can be purchased at 2s. 6d. a half-hundred-weight block.   Potatoes and onions at 9s. a case.

This enterprise of Sir Alfred Jones has already developed into the creation of cold storage companies at ports like Lagos, Calabar and Seccondee, and the firm of Elder Dempster has now built chambers on some of their ships capable of carrying twenty tons of European provisions every week to Seccondee alone.   The health of West Africa, bad though it is, has greatly improved within recent years, and though, of course, the medical profession has so largely contributed to the change, the house-builder, the merchant and the ship-owner have loyally co-operated in an endeavour to lighten the burden of the white man in West Africa.

III

GOVERNMENTS AND COMMERCE

NOTHING in West Africa is more striking than the attitude adopted by the several colonizing Powers towards commerce. At present, Germany is easily in the front rank ; her policy towards business men is the most enlightened of any Power, and it is therefore to be the more regretted that her treatment of the natives is not equally far-sighted. Were it so, all students of African questions could view with equanimity her gradual absorption of the whole of Equatorial Africa.

The British merchant knows with absolute certainty that he may rely on receiving a warm welcome and every assistance in German colonies. He knows, too, that none will be given a preference before him. He knows that if " public good "—the stick which governors so frequently wield—demands the removal of his factory, or that a road must be driven through his ground, the German Government will not quibble over doubtful legal points, but will look at the question on broad lines of common-sense policy.

Steam into a German port, and before you cast anchor you may see the customs and health-officers with their launches racing across the intervening stretch of sea. Promptly and smartly the doctor steps up the companion-way, and you begin unloading your cargo without further formalities. Your cargo finished, there is no delay about papers, no irritating objections about the closing time of the customs, or the doctor being at dinner or more likely, tennis. Contrast this with a visit to a French or Portuguese port—you may wait an hour before the health-officer comes on board. His visit over, the ship's officers and native crew slave throughout the day to unload the cargo, so that they may have the valuable night watches for steaming to the next port, but if the Frenchman can by any quibble keep you tossing at anchor, you may rely upon his doing so.

The German neither likes nor dislikes the British merchant : he is concerned with one thing only—that British capital and British brains are good for his colony ; therefore, without any sentimental nonsense, he gives the Britisher a warm welcome, and sees to it that no preference is given to the German merchant, which might make the British firm hesitate to invest further capital in a German colony.

Of course the regulations in German colonies are numerous and enforced with military precision and sternness. The native, centuries behind the white man, does not bear the strain very well. The Britisher, after a time, learns that such regulations are for his good and accepts them. No merchant at first takes kindly to keeping his back-yard free from refuse ; if he is in Togoland he resents the first instance upon which he is fined twenty marks for leaving old tins, half-filled up with rain water, lying about the rear of his store, but when in the process of time he is still without fever, he sees the advantage of this anti-mosquito regulation.

In Lome the Germans have an extremely interesting and unique system of transport enterprise. The surf, as in many parts of West Africa, is extremely bad, and for years constituted a source of perpetual loss, not only of valuable cargoes, but of human life. With characteristic thoroughness the German, at great cost, ran a pier out to sea, built a railway line on it and extended this line along the front of the merchant houses—a distance of about 1½ to 2 miles. On the pier the Government erected seven powerful steam cranes. Having laid down this plant, they took the next truly Teutonic step and compelled all the merchants to accept Government transport.

An outward-bound steamer is sighted at sea, cranes are prepared, the health-officer leaves before the ship comes to anchor, papers are examined, cargo is rapidly placed in the surf boats which are towed across to the pier where, in an almost incredibly short space of time, fifty tons of cargo are hauled up on to the pier, put on the train and delivered at the merchants' doors. A similar method is adopted with a steamer from the south—homeward bound. The moment the look-out ascertains her name and destination, he signals or telephones to the merchants, and shortly afterwards trains are in motion collecting the cargo already prepared for the expected vessel. When she comes to anchor, her surf boats are despatched to the pier, where they are promptly loaded and sent back to the ship.

There is a scientific air about the whole transaction ;

an absence of fuss ; an attention to business quite refreshing in tropical Africa, and above all, there is a sort
of " hey presto " promptness in the way these tons of
pots and pans, bales of cotton, barrels of oil and bags
of corn are handled.

All merchants, of whatever nationality, must accept
this transport and pay a fixed rate of 11s. a ton, which
covers all costs and insurance against every risk. In
return they are saved the expense and trouble which
attaches to the up-keep of boats, boat-boys and a large
staff of men for handling cargo. I was assured by the
merchants that the system works extremely well, saves
them much annoyance, and, on the whole, does not work
out at much greater expense than the rough-and-ready
methods of other colonial ports.

The administration of German colonies is decidedly
autocratic, although not more so than in British Crown
colonies. In German Cameroons, however, all interests
are consulted in a manner which demonstrates the
eagerness of the German Government to keep on good
terms with the merchant. Twice, sometimes three times
a year, the Governor holds an enlarged " Colonial
Council," to the deliberations of which he invites not
only the principal merchants, but the leading missionaries.
I was informed that at these meetings the Governor
welcomed criticism of existing or projected enactments,
no matter from what quarter they came, and that the
result was that everyone felt himself to be an integral
part of the colony.

How different the French Administration ! The
Entente Cordiale may be all right in the Banqueting Hall,

and as a pin-prick for Germany, but it is time the British people questioned its value in things that count. The truth is that in French colonies, merchants of other nationality are not wanted. Wherever you go in French West Africa, the merchant is full of grievances with regard to the petty annoyances of the Government and the officials. Nor does this apply to West Africa alone ; the same story is told in Madagascar and the New Hebrides, in both of which places, not only is the merchant entirely *de trop*, but the *Entente Cordiale* has not even secured decent treatment for the devoted missionaries. The *Entente Cordiale* was not brought about for selfish ends by Great Britain, and considering the much advertised generosity of our partner, we have a right to expect at least ordinary civilities in her colonies. The French are so absorbed in themselves that they would have none but Frenchmen on the face of the earth. As Napoleon failed to accomplish this end, the present-day Frenchman will not, if he can help it, have any but his own nationality in French colonies.

The Portuguese want British capital, but they don't want British merchants ; they kill the commerce of British firms by every form of preferential treatment. Their right to do so is, of course, equal to that of a man to cut his own throat. The only British enterprises in Portuguese West Africa are the Lobito-Katanga Railway, the Angola Coaling Company and some electrical works at Catumbella. The first named is the well-known Robert Williams' project for reaching the Katanga and Northern Rhodesia from the West Coast. The local Portuguese would probably like to strangle this valuable undertaking in its infancy, but they see already how much capital is finding its way into Angola. When Robert Williams gets his railway through to Katanga, the Angola colony will become an asset of considerable value to the Republic.

The attitude of the Belgian Government towards commerce is again different from that of any other colonial administration. Theoretically, the Belgians are anxious to persuade capital to enter the colony, but the principles of King Leopold's rule have taken such firm root that in practice the presence of any commercial agents, particularly those of any other nationality, is gall and wormwood to the local Belgians. Nothing, for example, irritates them so much as a reminder that by the Berlin Act they are bound to keep the country open to the free commerce of the world.

Even Belgian merchants complain of the treatment they receive at the hands of the officials of the administration. Recently, when calling at Stanley Pool on board a merchant steamer, we had to pass the customs official. We put our anchor ashore in front of the customs house, where the official himself was standing on the beach smoking a cigar, and, as we thought, waiting to examine our papers. He knew the captain (a Belgian) was pressed for time, yet he deliberately kept the ship at anchor for twenty minutes whilst he finished his cigar ! No doubt this conduct was meant to—and, of course, did—impress the crew, but, as the captain remarked, the reason at the back of such action is the desire of Belgian officialdom to monopolize transport, and their hatred of any form of free commerce.

I was present on another occasion which instanced Belgian desire to secure trade in principle, whilst unwilling to put their advertised desires into practice by exhibiting a readiness to render real assistance. There came into Boma a British ship, whose captain was of higher rank than those usually visiting this port; it was in fact the first time this officer had called at a port so insignificant as Boma. He ran his ship alongside the pier, but was amazed to find none of the ordinary preparations for unloading cargo. Instead of sending a ship's officer for an explanation, he went himself to see the quasi-Government Railway Company.

"Where," he asked, "are the railway trucks for unloading cargo?"

"There they are," laconically replied the official.

"But I want them at the ship," said the captain.

"Well," answered the official, with genuine courtesy, "you can take them, I don't object."

That it was in any sense the man's responsibility to send these trucks along did not occur to him, and upon the captain asking how he was to get them over the intervening half mile of line to the pier, he was told, again with every courtesy, "send your crew to push them!"

Then might be seen the spectacle of a ship's officer and a gang of Kroo boys spending hours under a tropical sun straining and tugging at these unwieldy railway trucks, all of which could have been shunted in a few minutes with ease by any one of the idle engines in the sheds. That a ship of 5000 tons was delayed for twenty-four hours by this stupidity was immaterial to the Belgian official. How differently the German would have acted! The empty trucks would have been ready on the pier, a shunting engine with steam up standing by directly the steamer began making her way alongside, but the Belgian is not cast in that mould.

In British West African colonies the relations between Government and Commerce are unique. Alone among the Powers she has developed a caste attitude, until to-day the distinction is not a little embarrassing. The British official is quite a good fellow when you get him alone, but, as a class, they form a distinctly objectionable " set." This is apparent the first day on board ship, when the " sorting out " commences, and if the weather is good this process provides not a little amusement to an observant passenger. Usually there are but three groups of travellers on a " coast " steamer—the official, the merchant and the missionary. As we have travelled a good deal in these ships, many occasions have presented themselves for watching the arranging and re-arranging of this little floating town. The last time we set out from Liverpool was the most entertaining of any. Running down the channel, a youth, who had apparently never travelled before, wished me " Good day," with the apparent intention of pacing the deck, but upon his discovering that I was neither an official, nor a missionary, he inwardly argued " a trader," and promptly made off!

Another and yet another pursued the same tactics, until by a process of elimination they " discovered " the officials. " Steward "was then called and all the "official chairs " were placed in a semi-circle in the best part of the deck. That this monopolized the only comfortable

CATARACT REGION BELOW STANLEY POOL, BELGIAN CONGO.

section of the upper deck did not appear to concern these gentlemanly youths.

In the dining-saloon the chief steward had placed us at one of the lower tables, but learning from the captain of certain instructions given him by one of the Directors, with whom I was on friendly terms, this man came forward and with profuse apologies asked me to accept an entirely different place in the saloon, saying that he " thought I was a trader ! "

Once I met a young Sierra Leone merchant, who told me that a certain official in the Protectorate had been taken ill with a bad fever at his factory ; that he had nursed him through it with all the care of a relative ; that this official, when he was at last able to leave, appeared deeply grateful for all that had been done for him, and the merchant believed he had made a lifelong friend. A few months afterwards business called him to Freetown, and passing along one of the streets, he met two or three officials, one of whom was the friend whom he had so carefully nursed. To his amazement, he only received a curt nod and a plain intimation that further intercourse was undesirable. It is to be hoped that such conduct is rare, but the general attitude of the younger British officials is becoming almost intolerable.

This treatment of the merchant class finds no place in any other colony of West Africa. It is of quite recent growth and monstrously unjust to the merchants, for it should never be forgotten that it is almost entirely to the merchant and missionary communities that Great Britain primarily owes her presence in West Africa. There is another fact our officials would do well to remember, namely, that the natives and the merchants together pay their salaries and pensions.

The younger officials make themselves far more objectionable than the older men, but probably this is due to their inexperience. It is, however, regrettable that the older officials do not set a more pronounced example in the other direction. Within recent years, the British Colonial Office has been sending out, in the capacity of Assistant District Commissioners, many youths of necessarily immature judgment and totally lacking in experience. These lads are by far the worst specimens in their attitude towards the native and merchant communities. Recently, this feature has been impressing itself upon travellers in East as well as in West Africa. Mr. E. N. Bennet, in his book on the Turks in Tripoli, says :—

"Amongst our fellow passengers to Marseilles
"were eight young men who were on their way to
"Uganda. Few, if any of them, had ever crossed
"the Channel before ; they wore school colours and
"did not know an olive tree when they saw one.
"Nevertheless, they held, and expressed, very
"decided views—the ideas of the College Debating
"Society and the London Club—that the 'man on
"the spot' must be the sole arbiter on matters
"colonial and that kindness was absolutely wasted
"on black men ; the one ethical quality necessary
"in a representative of Great Britain was firmness.
". . . They also viewed with disfavour the deporta-
"tion of Mr. Galbraith Cole. One could only hope
"that when these inexperienced youths grew older

" they would grow wiser. As it is, an immense
" amount of harm is done all over our vast Empire
" by some of our younger soldiers and civil servants,
" who, utterly devoid of *cosmopolitanisme gracieux,*
" treat their non-English fellow subjects with a
" contempt which would be ridiculous if it were not
" dangerous."

The merchant seeking a new field for commerce in
West Africa will find the warmest welcome and the fairest
treatment in German colonies, and next to Germany, in
this respect, the British colonies ; there is not much to
choose between the Belgian and the Portuguese. None
but Frenchmen should go to the colonies of " Liberty,
Equality and Fraternity," for there is little Liberty,
less Equality and no Fraternity in the French colonies
for white or black.

IV

THE LIQUOR TRAFFIC

IT is useless to close our eyes to the fact that an evil of fearful potentiality is being introduced and fostered all down the West Coast of Africa. I have not always found it possible to agree with the much-criticized Native Races and Liquor Traffic United Committee, but it must not be overlooked that some of their critics have made errors, in judgment at least, not one whit less extraordinary than those which have been brought against that Committee of highminded and unselfish men.

The greatest mistake made by people in Europe upon this question is that of comparing it with the European consumption of alcohol. The African is not a drunkard in his primitive state and he detests our ardent spirits; once in an extremity I gave a young man a sip of brandy in water from my medicine case, and he literally howled over it and set his teeth firmly against my trying to give him another dose !

The error to which most people cling so tenaciously is that of the "scoundrelly merchant" theory. They cannot understand—because they do not know Africa—why a merchant should pour gin into West Africa, unless he is making a fortune out of it. As a plain matter of fact the merchant makes less out of the sale of alcohol than he would out of almost any other article of commerce. In a village store on the Gold Coast hinterland, I found rum costing 6s. 9d. a gallon being retailed at 7s. 3d.—a profit of only 6d. per gallon. In another store I visited, the native merchant was retailing gin at 9d. a bottle, for which he was paying 8s. 1½d. per dozen, and 4d. a case for transport to his store. A West African merchant once remarked to me, " If you could stop the demand for intoxicating liquor it would pay me to give you twenty thousand pounds." The merchant was quite right, because, whilst he could get fifteen and twenty per cent. on the sale of Manchester cotton goods, he was only making a few pence a case on the gin he was shipping to Lagos ! The sale of alcohol does not pay the merchant, but we cannot escape from the fact that it is a good revenue producer.

There seems to be a general impression that the British administrations are the worst in this respect, and that their record is not without fault few would deny, but I am confident that the moral sentiment of the British Government and people will save them from falling so low as the French administration—an easy first in almost all that is retrograde in Equatorial Africa. France to-day recognizes the terrible evils which follow in the train of Absinthe-drinking in the homeland, yet she can calmly look on whilst natives stream into the little drink stores of French Congo with their 25 cent pieces to purchase " nips " of what I was assured by the vendor was the worst form of drink in the whole of the African continent. When we were at Gaboon, an official informed me that quite recently two young

Europeans had taken to drinking trade Absinthe, and in each case had died in a manner which called for a post-mortem examination, the results of which horrified the examining doctors.

The Portuguese have long been regarded as by far the worst sinners, but it is the fashion in West Africa to place every sin at the door of that not unkindly nation, yet however deeply they may have sinned in the past, there are happily signs of repentance and reform. In Angola the Government has recently decreed the abolition of distilleries throughout the colony, providing, out of their extreme poverty, considerable sums as compensation for the manufacturers.

The Belgians lead the way among the colonizing nations in West Africa, for in their colony they are bringing the prohibition line ever nearer the coast and it is now impossible even in the " open " areas for a native to purchase any intoxicating liquor between Friday night and Monday morning.

If the natives as a rule dislike alcohol, if the natives of West Africa are less drunken than Europeans, what happens to this ceaseless and increasing flow of spirits into the West African colonies ? " Over one million cases of Hamburg spirit are retailed to the natives here by a single firm within a year." Such was the remark passed by a dispassionate Government official to me when in Southern Nigeria. There are twenty or thirty big merchants in Lagos alone, who handle huge consignments of this spirit by every steamer. Sitting on the banks of the Lagoon, one sees an endless stream of small craft passing to and fro with their loads of gin,

going to a hundred different centres, some with only six
cases, others with fifty and even one hundred. I visited
a farmer up country, who admitted to me that he retailed
over £1000 worth of gin and rum every year. The
same story met us at Abeokuta, where something like
thirty-three per cent. of the imports are spirituous
liquors, and the returns published show that in the month
of January, 1911, out of a customs revenue of £2644, no
less than £2450 came from duty on spirits.

None deny, because they cannot, this prodigious
importation of spirits into the Gold Coast and Southern
Nigerian territories ; but one thing baffles every observer
—where does it go ? The Egba and Yoruba people of
Southern Nigeria are not drunken. We could find very
few white people who had seen any appreciable degree
of drunkenness ; generally it was suggested that drinking
took place at night. In order to test this theory, I went
several times, at a late hour, quietly through the lowest
parts of Lagos town. I saw many things, some of an
appalling nature, but no single drunken man or woman
could I find, and the statistics for convictions barely
show one per thousand of the population.

Yet we cannot escape from the official figures. Over
six and a half million gallons of spirituous liquor of
European manufacture were imported last year into the
British colonies of Sierra Leone, Nigeria and the Gold
Coast.

What happens to this increasing stream of spirits ?
No one has ever been able to give a satisfactory answer
to the question. Some say that being a currency,
millions of bottles of gin are " banked," i.e., stored ;

some say that large quantities are consumed at festivals ; others assert that it disappears in secret drinking. I am inclined to think, however, from visits paid at all hours to the people's homes, that spirit drinking is spread over a much wider area than has hitherto been thought ; that is to say, moderate drinking prevails widely, but that at present few of the natives drink to excess. If the moderate drinking of to-day is leading the people to drunkenness to-morrow, then a catastrophe of first magnitude will fall upon West -Africa. Drunkenness is admittedly on the increase in the Gold Coast, and this is so obvious that three years ago the Governor sounded a warning by saying that he recognized drunkenness was becoming one of the most dangerous enemies to Christianity.

What is to be done ? Everyone admits that the sale of intoxicating liquor to natives (many would also add—and to whites) in Africa is an evil ; all are agreed that the danger is potential rather than actual. But very few seem to have any other remedy than—repression, prohibition, high licenses, heavy duties ; these are the methods which find greatest favour to-day.

Prohibition is an extremely difficult proposition for any African colony, and it is well-nigh impossible where the French and German boundary lines march with that of another colony. If, for example, Great Britain proclaimed prohibition for the Gold Coast, what guarantee have we that German native traders would not smuggle spirits across the Volta into the Gold Coast, or the French traders carry it over the Dahomean border into Southern Nigeria ?

High license and import duties have both been tried, and both failed to check the growth of imports. In some places, it would seem that these very restrictions make matters worse. I was informed by a white doctor on the Gold Coast that chiefs in the hinterland will take out a license sometimes of £50, or even higher value, but will impose a tax of 5s., or more, per head, on the entire community to pay for it. My medical friend, who was a man of long experience and wide knowledge, further said that many of the people resented this tax because they were abstainers, and on that ground complained to the District Commissioner, but the only redress they obtained was, " Call it a loyalty tax then, and pay it ! "

It would be interesting to see what would happen if the duty as a prohibitive measure were temporarily removed. I do not think it is altogether clear that it would tend to increase the consumption ; one thing is certain, it would cause something like a financial panic amongst those natives who, holding large stores, hope that the agitation in Europe will enhance the local price and thus make possible extremely profitable sales of stocks.

There are two spheres of action entirely untouched to-day. West Africa is a very " dry " place indeed, and the thirsty inhabitants must have some beverage other than water. Palm wine used to be the national beverage, but the demand by Europe for the products of the oil palm is so great that the whole strength of the tree is required for producing vegetable oil.

The other sphere of operation is beyond question the

most effective—an internal movement against the consumption of, and trade in, spirits. Repressive measures
by Governments are all very well in their place, but
without the goodwill of the people those measures cannot
be wholly effective. An agitation locally kept up with
the vigour that characterizes the campaign in England,
would do an enormous amount of good.

For generations past we have been telling the native
that he, in his primitive state, is everything that is bad.
Certainly the African, modelled upon a combination of
the reports of travellers, officials and missionaries, is a
creature the devil himself would disown. Unfortunately,
the native has, to some extent, come to believe this, and,
abandoning his native rôle, has struggled to imitate the
whites who, he has been taught to believe, are the highest
type of civilization. When, therefore, the white man
ships his gin to the African, he considers it the " correct
form " of the higher civilization to purchase it, and copy
the European to the extent of drinking " gin and bitters,"
" gin and water," " whisky and soda," " cocktails " and
other liver petrifying abominations, forsaking his simple
draught of water and his kola nuts for the drinks that
help him up to the standard of his inexorable critics and
overlords.

The Governor and his officials can, if they like, do
more to stop spirit-drinking than all the prohibitions,
taxations and high licenses that the wit of man could
impose. Is it impossible for one colony to set an
example ? I think not, for I believe the British officials,
as a whole, in spite of their shortcomings, are capable of
making any sacrifice for the good of the colonies. If a

governor would " set the fashion " and by his example inspire his subordinate officers with a determination to refuse to drink any intoxicating liquors in public, at any function or ceremony whatever, for a period of three years, and thereby set the fashion against spirit-drinking, I venture to predict that within those three years the import of spirits would decrease by at least one half. The natives, rightly led by the Press, and the movement supported by the officials and by the ministers of the native churches, would take fire, so to speak, until the drinking of spirits would become " incorrect form."

In the hands of the Government officials is the power to turn the natives by example against the consumption of ardent European spirituous liquors. Will they seize the opportunity ?

V

THE EDUCATED NATIVE

THE man who would understand the African must get beneath the surface, otherwise he will never know the real sentiments of the native races. By confining himself to the hospitality of the whites, he will learn a great deal about the natives, and will also learn to appreciate the position of the merchant and the administrator, but if he would probe the mind and thought of the African, he will find no better way than that of living with him.

It is of course more congenial—to many essential—to accept the hospitality of trader or official, for there are little things a native host and hostess will inevitably forget ; but the compensations ! What a wealth of affection, courtesy and native lore is poured at the feet of the visitor.

Driven by fierce tornadoes, wet, cold and utterly miserable, I have sought the simple hut of the forest hunter, or the fishing-shed on the banks of an African river. How warm the welcome ! How quickly the good wife will bring forward native refreshment ! Let a drop of rain find its way through the roof into the hut and on to the white guest, and nothing will stop the impetuous host from dashing outside in the foulest of weather to stop the leakage. Readily, too, he gives up his rough bed and will curl up in the hollow of a tree, or beneath its branches, joyfully enduring any discomfort so long as the white man may be made comfortable.

It is the same at the other end of the scale. Those who discover that terrible disease—negrophobia—creeping over them, often in spite of the better self, will find an infallible cure by staying for a few days with some leading educated native. Their view-point will almost unconsciously change under the genial and enlightened conversation of the dinner-table ; their hostility will melt away under the influence of the natural courtesy of the warmhearted host. They will begin to marvel that some things should never have occurred to them before, and, unless race prejudice closes the observant mind to all reason, the guest will forget that his host is an " accursed educated African."

The " educated negro " is to many only a worse evil than the primitive savage, but what has the educated native done ? What terrible crime has he committed ? I admit he has imbibed the education European civilization provides, but is that a crime ? I admit that he is probably a greater consumer of spirituous liquor than the illiterate native, but if it is wrong for the native to follow in the footsteps of his white exemplars, why does the white man import it ? I admit that he is often over-dressed in too demonstrative European clothes, but again, if it is wrong for him to wear these things, why does European compete with European in producing the liveliest patterns in clothes and the most outrageous collars and boots ? If these are the things which make the educated native unfit to live, why send them to him ?

I am not here concerned in condemning the sale of European outfits, importation of spirits, least of all European education, but in fairness to the African, let us brush aside unreasoning and unreasonable prejudice and put ourselves in his place for a moment. Let us at least recognize for example that if grave faults exist in the educational systems we provide for Africa, it is upon us, rather than upon the African, that the responsibility rests.

We all agree that the educated African has his weaknesses, and pretty bad ones too, but though I have met hundreds of them, though I have read volumes of material they have written, I have never met one who claims the perfection in life and conduct that not a few of his critics assume. It seems to be mainly in British colonies that the educated native is such a bugbear, and if our educational system produces such evils, it is done after all under an autocratic and not a representative Government. Surely, therefore, we should lose no time in abolishing, root and branch, the cause of the mischief.

But is it a failure? If so, wherein has the African failed? Take first the elementary curriculum of mission and Government schools. Where would Africa be to-day without its thousands of coloured clerks and Government officials? In Southern Nigeria alone there are 5000 natives in the British Government service, all of them more or less educated. In every colony, too, you meet cultured natives trained at these schools who are now devoting their lives to the education of the rising generation.

We are told that the education of the African has

been too largely concentrated on a purely literary and spiritual curriculum. Beyond question there is some force in this criticism, but the Missions and Governments are surely more responsible for this than the natives themselves. The Government particularly so, for missionary committees are after all only trustees for the funds placed at their disposal, and such are almost entirely given for purely missionary propaganda. But even this criticism is unjust in ignoring the existence all over West Central Africa of the educated native carpenters, bricklayers, engineers on steamers, engine drivers and guards on railway trains.

Crossing the Kasai territory I met an American Bishop, who had also travelled not a few thousand miles in Central Africa, and this charming old Divine could not cease exclaiming, " Well, the way you English are covering this continent with educated native carpenters, bricklayers and engineers is just marvellous." Go where you will, you meet these men. In the upland cocoa roças of San Thomé, in the workshops of German Cameroons, in the trading factories of almost every island, you will always find the Accra or Sierra Leone trader and mechanic who has received a fairly liberal general education at the mission schools. A thousand miles north in the Congo, away south towards Rhodesia, you will hear frequently the welcome salutation, " How do you do, sir ! " Welcome then indeed is the claim to one Throne one Empire ; more welcome still is the kindly assistance with baggage, the clean hut, the generous gifts of fruit and provisions.

" May I pay you for your kindness ? "

" No, sir, I am too glad to see you. God bless you, sir. Goodbye."

The traveller thus refreshed goes on his way and vows that when he gets home he will send a subscription to those missionary societies who are sending forth this stream of men to the distant parts of the dark continent.

The principal openings for the sons of native chiefs are the medical and legal professions. First let it be remembered that the enlightened chiefs fortunately saw that by giving the flower of the race scientific European education the power of the witch doctor, who, throughout African history has been both medical and legal quack, would be broken. Not only so, but the sick and afflicted among the race would receive the best alleviation that science could provide.

Has the coloured barrister failed? If so where? Certainly not in British examinations where brains and energy provide the only standard. I shall probably be told by the critic that he has failed in practice. If this be so, how is it that whenever a Crown case comes along the British Government promptly briefs leading native barristers?

Has the doctor failed? Again, where? Not in the English and Scotch hospitals, for he has frequently carried a higher degree than he finds amongst his European colleagues when he returns to the coast. That he is excluded from Government service proves nothing, except perhaps prejudice. It may be asked why in the Gold Coast colony the African medical man is allowed no place in Government service. We are told in reply,

DR. SAPARA OF LAGOS, A MEDICAL MAN IN THE SERVICE OF THE
BRITISH GOVERNMENT.

(Dr. Sapara is endeavouring to persuade the natives to adopt attire more suited to Tropical Africa
than the frock coat and silk hat of the European.)

because white men, and more particularly their wives, would refuse to receive treatment at the hands of coloured medical men. This argument fails entirely when we remember that the majority of the hospital patients are not white, but coloured, and at present can only receive treatment from white doctors. Moreover, do we not know of white men, who, fearful of that rising temperature, that throbbing pulse, unable any longer to bear the suspense, have sent for a native medical attendant, and under his kindly treatment have recovered, in some cases to remember gladly the skill exhibited, but in others, alas, too easily to forget that they owe their lives to such tender ministrations.

Then, too, are there not to-day many white men on the coast who prefer native doctors—whose names I could mention—to the services of European medical men ? Have we not heard and known of something still more eloquent—the calling in of native medical men to white women ? Many a white merchant and Government official has taken out a delicate and highly-strung wife to assist him in his work, and almost every " coaster " knows how one of these heroic women was stretched upon, apparently, the last bed of sickness ; the distracted husband had tried everything, had implored the white doctor to try something—it hardly mattered what—to give back health to the sufferer. Suddenly a thought occurred to him ! The native doctor, fully qualified, was sent for and visited the patient, and then in consultation with his white colleague, other treatment was tried. Slowly the sick one fought her way back to life and health, and to this day the husband remembers to whom he owes the restoration of one who to him was everything—and this is no isolated case.

When death's angel looks in at the window, which is pretty often in West Africa, race prejudice shamefacedly slinks out through the nearest doorway.

The administrator, the missionary, and the native, however, realize that the educational facilities at present at the disposal of the natives are not ideal ; the march of progress has shown defects, and these must be remedied. If there is one administrative problem in British colonies important above another, surely it is that of education. In all things colonial, Great Britain has hitherto given a lead ; let her maintain that proud tradition by appointing a commission to study the whole question of the education of the African peoples in her Equatorial possessions, with the object of ascertaining how far the Government may be able to secure a more even balance between the literary and technical training of the natives ; how far it may be possible to so re-adjust existing systems as to avoid denationalization ; how far it may be possible to extend that supremely important but largely neglected branch of education—practical agriculture.

Then there is the question of Government grants. Can anyone defend the antiquated system which prevails in many colonies of giving lump sums of revenue to missions ? An excellent departure from this rule has been commenced in the Gold Coast, whereby the missions receive a grant per capita for the finished product, *i.e.* when a scholar reaches a given standard in literary and technical knowledge, the Government makes a definite grant of from 20s. to 27s. 6d. for each scholar attaining to that standard. This experiment, already fruitful of so much good, might provide a model for other parts of the African continent. A commission could study how far this should be extended and whether it might be wise to lead on to scholarships for an extension of the education by providing grants for the study of agriculture in the botanical gardens and plantations of the tropical world. For example, if facilities were provided certain natives from the Gold Coast would derive great benefit from a study of cocoa plantations in different parts of the British Empire.

If race prejudice were too strong to admit of this procedure within the Empire, then such natives would undoubtedly benefit by a visit to the plantations of other Powers, in particular those of the Portuguese on San Thomé, where, although there is predial slavery, no race prejudice exists which would prevent a close study of one of the finest systems of cocoa production in the world —certainly second to none in West Africa.

Another problem which knocks loudly at the door of the British Colonial Office for consideration is that of Africans seeking a legal and medical education in the Mother Country. We cannot, and have no right to object to their doing so ; on the contrary, we ought to welcome the idea, to be proud of the fact that our Administrations are so progressive that they help this movement forward. But we are not ; we do not like some of the results which at present attend this practice. Again, I ask, are we not responsible ? These young men at a most receptive age come in all their enthusiasm to the Motherland of their dreams ; they expect to find a civilization, but one remove from the realms of eternal purity and bliss, and what do they find ?   No strong and friendly hand is outstretched to help them, no responsible person comes forward to take them by the hand and bring them in touch with the better elements of our national life.   Alone in London or Edinburgh they drift into the worst channels and imbibe the most pernicious ideas and practices that float around the parks and parade themselves in the streets of our great cities.   What wonder that their lives are fouled ?   Who can be surprised if the only seeds they carry back to the colonies are those evil ones which produce a crop of tares to the embarrassment of the Government ?

Philanthropy can do much to turn the thoughts of these young men into loftier channels, but philanthropy should not be left to do this work alone.   Surely the Colonial Office, if it has no duty in the matter, at least for its own sake could render some assistance in giving these young students a closer knowledge of the men, the aims and the desires that inspire British Adminis-tration.   In the whole world there is collectively no finer group of officials than those in the service of Downing Street ; some seem to think they too closely resemble highly-specialized machinery ; some of us know otherwise ; some of us know that behind the official mask there are men whose hearts and consciences pulsate with lofty principle and humanitarian sentiment.   Yet between this wealth of goodwill and experience, and the African youth amongst us, a great gulf is fixed ; there is no medium of friendly intercourse between these noble-minded officials and ex-officials of the Government and the young Africans who are being trained to mould the character of their compatriots and of public opinion in Britain across the seas.

John Bull must wake up to the existence and the needs of these children, must realize that their education, whether in the colony or in the Mother Country, is of supreme importance, and that the friendly and wise oversight of their education is an Imperial responsibility of the highest order. It is more, for all nations have looked to us in the past for the solution of these problems, and upon such facts—rather than upon a colossal navy— rests the real strength of Great Britain.

VI

JUSTICE AND THE AFRICAN

THE Powers of Europe—and Great Britain in particular—
boast of the " justice " with which they treat native races.
Happily the native tribes, as a whole, fully share this
complacent belief in European rule, and this no doubt
arises from the fact that before the Powers of Europe
divided Central Africa between them, justice, as compared
with might, had but a small place.

This belief, however, is perceptibly passing away,
and in many of the West African colonies the natives
are not now prepared to accept, without question, the
acts of European administration. To such an extent
has this feeling grown within recent years that administrative action sincerely taken in the best interests of the
natives is frequently assailed.

No one would deny that blunders are but human ;
few would deny that the finest Colonial Office in the
world—that of Great Britain—has made mistakes which
subsequent history condemns. The natives have enough
common-sense to make every allowance for such mistakes,
but what they do not understand—and in this they are by
no means alone—is, why recognition of the mistake is
not made promptly, and some reparation made for the
error. The plain man asks why there should be some

Medo-Persian law which forbids the admission of error and the consequent refusal of reparation. This attitude is accountable for much harm to the prestige of the European in West Africa.

In Government despatches, in speeches, in our schools and from our pulpits, we are never tired of preaching upon those articles of British political faith which know no party. We pride ourselves upon our love of justice and freedom, and yet we do things which we know to be utterly indefensible, which we know to be in entire contradiction to our belauded principles. We have made the blunder and we know it, but we invariably crown it with the further blunder of refusing to admit it.

We know perfectly well that it is indefensible to arrest a man and arbitrarily punish him without trial, but it is done, nevertheless. During our journeys in Central Africa, we visited a grey-haired old chieftain living in a hut on the Gold Coast. The old man was reclining in a cheap deck-chair, he was totally blind and unable to stand. What was his story?

Some thirteen years ago he heard rumours of a rebellion against the British Government in Sierra Leone, and immediately Bai Sherboro sent a message to the District Commissioner that the war boys were bent on attacking Bonthe. This timely information permitted of measures being taken to protect Bonthe. One day a messenger called upon Bai Sherboro and told him the Governor wished to see him. Trustingly the old man picked up his staff and went to the British authorities, when, without trial—and he asserts without being even informed of the charges made against him—he was forthwith exiled to a lonely spot on the shores of the distant British colony, the Gold Coast. In Sierra Leone the old man had a son, who, refusing to allow his father to go forth alone, sold all he had and joined him in solitary exile, and who to this day shares his loneliness and sorrow.

The British Government does not deny the facts, but, apparently acting upon the advice of the " man on the spot," who has probably never seen any other person than the old chief's interested accuser, takes up the position that the return of this blind and decrepit old man to his native home, would be dangerous, on the ground that he was believed to be implicated in a rebellion ! The Government has all along refused to give the old man a trial, so that he might face his accusers and meet the charges, with the result that he must die in exile. There is something very un-English about such an incident. Strangely enough the old man still holds firmly, after all these years, to his admiration of British rule, and faith in British justice. Again and again he reiterated to us the words, " If only the King of England knew ! " " If only the King of England knew ! "

This is a passionate loyalty which surely we are unwise to trifle with, unwise to immolate upon the altar of theoretic administrative infallibility. It is folly to bury our heads in the sand so that we may not see these things, for if we fail to look these facts squarely in the face, others are regarding them—our friends with deep concern, our enemies with the keen relish of an insatiable hatred.

Will it be argued that this is only an incident ?

Possibly, but who knows ? This case was unknown to the outside public until the old man's hair had whitened and until he had lost the use of his limbs during ten years' exile. Two years of persistent knocking at the door of the Colonial Office has even failed to secure permission for the old man to return to die in his own country.

Many chiefs and native merchants in British West Africa have but one ideal for their offspring—to send them to England for an education either for the bar or for the medical service. They are pathetic stories which some of these men tell you of how they deny themselves and their families so that they may save enough to send " my eldest " to England. They themselves have only heard of the glories of England, they can never hope to see them, but their determination is that the boy shall. The latter comes and spends his four or five years here in England, possibly more, and during that period the old man is slaving away on his farm, or trading early and late in his store, has watched his savings trickle away until often he has but little left. At last the glad day of home-coming arrives. The lad steps ashore from the boat, a fully fledged " medico," carrying " no end of big degrees." How proud the father is ! How amply repaid he feels for all his efforts and struggles, as his full-grown son explains to him the degrees he has obtained are higher than those of Dr. Smith, the white medical officer at the hospital.

The young medical man hopefully sends in his request for an appointment in the Government service—an appointment which must be paid largely from native

taxation. At a later date he receives an official envelope,
which he greedily tears open in the presence of the
expectant and admiring family. It is the official form,
intimating that his services are not wanted !

We all know the reason, why wrap it up in gentle
phraseology, the hideous fact is there—the medical
service is the monopoly of the whites. Of what avail
are degrees of the highest order ?   What use is it to argue
that native medical officers would be less costly ?   The
colour-bar is thrown across the threshold of opportunity
in the Gold Coast. The young man himself understands,
possibly he may even come to hate the Administration
which appears to hate him, and can we be altogether
surprised ? The old father does not understand it, he
is bewildered—the blow that has fallen upon his hopes
is a heavy one, and in spite of himself he wonders what
is amiss with British justice.

The island of Lagos, measuring less than 600 square
miles, with a population of nearly 80,000, was always
congested, but never so badly as it is to-day. By day,
and also by night, I have traversed the native quarters
and found overcrowding which before long must produce
a grave condition in that hub of West Coast commercial
activity. Lagos is always hot, always humid, always
malodorous to epidemic point, but Lagos, overcrowded
though it was, has within recent years seriously added to
its congestion by the forcible expropriation of some
hundreds of people from the lands they occupied. No
doubt a nicely-laid-out race-course is more pleasing to
the eye of many British officials : the brightness and
neatness of this fenced park is cheering to those who now

have a monopoly of this vicinity, but the price paid for such expropriation is a further alienation of native loyalty and goodwill. Somehow the native does not like being driven from his home, even though " Hobson's " compensation is provided.

VII

RACE PREJUDICE

THE most lamentable feature which confronts the traveller in British West African colonies to-day is that with the growth of commerce on the one hand, and with the spread of Christian thought on the other, race prejudice is rapidly increasing its hold not only through an ever widening area, but in an intensity which must before many years have passed precipitate a grave condition in the relationship of the two races. The decks of West African liners provide an incomparable mirror for reflecting white opinion upon the shortcomings of the black man. On shore each man is busy with his own affairs and usually meets only men of his own circle, but on board ship one meets every class ; moreover, the conditions of travel tend to facilitate a flow of conversation. One sees stretched upon the deck, in every conceivable attitude of comfort and discomfort, all classes of the coast community : the dapper little colonel ; the young district commissioner ; the army doctor ; dealers in oil, ebony and rubber ; the Nimrod going out in search of big game, and the missionary going forth in quest of human souls. These varied interests cooped up on the decks under the enervating influence of the tropical sun will with some exceptions share little in common, but that of an indefinable dislike and contempt for that black man they come out to govern or exploit. To the student of human affairs, the conversation is of absorbing interest, revealing as it does every type of thought and superficiality. The loquacious trader, with the experience of but one term, opines with a lofty air that the " nigger " is the very embodiment of Satan. The " gentle " wife of Britain's representative suggests that the sum of all evils—the native we have half-educated, should be curbed by measures dear to the heart of the short-sighted statesmen of Russia. The sympathetic doctor, with ten years' practice, looks on and holds his peace, a silent but eloquent censure. The missionary, with longer experience still, likewise says nothing, but listens with pained interest. The deck below is filled with the usual crowd of natives : the tall Fulani trader ; the squat Gold Coaster ; the Christian servant from Freetown ; the devout Mohammedan merchant going up to Kano, possibly on to Mecca. The mammies, too, are there, dressed in skirts of brilliant Manchester print and gaily coloured blouses, outrageous in fit and style. The piccaninnies play their little games and romp round their admiring mammies. Not infrequently a child stands sadly apart, maybe a girl possessing but little in common with the other children, her little head with its pale face is covered with something half-wool, half-hair ; she has a father somewhere, possibly amongst that group on the upper deck, but between upper and lower deck a ladder is fixed, down which the white man may go whenever desire prompts him, but up which neither coloured nor quadroon may climb.

But what are these exceptional sins of the coloured

man ? What are these terrible shortcomings of which he has the absolute monopoly and which call forth bursts of passionate denunciation from the great men of the earth ? " An incurable kleptomaniac "—" unspeakably immoral "—" grossly impudent "—" incorrigibly lazy "— are but a few of the sweeping indictments hurled pell-mell at the reputation of the absent and mainly defence-less " prisoner in the dock." Civilization, which has never robbed the African of his land or its fruits, never bought and sold him, never violated his daughters, but has ever protected him, has ever set before him a perfect standard of Christian practice, should examine these whirling charges in the light of established facts. It cannot be denied that the African frequently breaks the eighth commandment, but there is some evidence that the Almighty had the Anglo-Saxon race in view rather than the African when He gave Moses the ten commandments on Sinai's mountain.

The following incident will show the prejudice to which the African is subjected : Our vessel was pitching, tossing and rolling her way down the West Coast, most of her passengers too sea-sick to stir far from the upper deck. A steward shuffled his way along endeavouring to balance cups of chicken-broth to tempt the appetite. One of the passengers helping himself, called attention to the lack of spoons. The steward replied : " We are not allowed to bring them, sir ; you see there's niggers aboard this ship ! " Though knowing perfectly well that the Kroo boy may not intrude himself upon the upper deck, even the steward seeks to make him responsible for losses more properly attributable to the members of his own staff.

The Post Office clerks at Sierra Leone, and Custom House officials at Lagos, are cited as paragons of impudence and " swelled head." It must be admitted that these men fully realize that they are servants of the British Crown and maintain a dignity not altogether appreciated by the white community. If they can be accused of " swelled head," may it not be that white example has led them to regard such an attitude as " correct form " for Government officials ? Examples of this may too often be seen in British Crown colonies, for between the British official class and the merchant community a great gulf is fixed, across which many officials gaze with unbecoming contempt. Let the subordinate native but ape this attitude, and, in him, it becomes a sin.

With bated breath and eloquent gesture, the frightful immorality of the native is a morsel of scandal dear to the heart of many superior whites. This is a matter, however, upon which students of African social life have some differences of opinion, but none have any such differences of opinion upon the necessity of " Form B," which so many white officials are prone to forget. An exposure of African immorality cannot, it is true, be long delayed ; sooner than most people think that day is coming. Locked in the breasts of governors, doctors, missionaries and educated natives are strange stories and appalling statistics ; their volume is daily increasing ; facts are being labelled and classified and these only await the opportunity which an increasing virulence of attack upon native immorality—ignoring that of the white race which obtains in every African town—will precipitate.

The chief indictment against the African is that of

being incurably lazy   .Prejudice has so blinded the eyes of critics that they do not see the fleets of sail and steam craft which the horny black hands send to and from the West Coast laden with produce. Look over a single ship ; there are boat-boys, deck-boys, boys for cleaning brass, washing plates and dishes, splicing ropes, hauling rigging and painting ironwork. " Boys " for loading barrels of oil, for towing and loading floats of giant timbers, all of whom, more or less, keep the doctor busy bandaging their crushed fingers and toes or sometimes their broken heads. " Boys," too, for delivering cargo ashore, through the wild surf in which many lose their lives every year.

Those who have a leaning towards the " lazy nigger " theory would do well to stand for a single hour at the Liverpool docks and watch that unbroken stream of drays heavily laden with tons upon tons of mahogany for our tables ; cocoa beans for our chocolates ; rubber for our motor cars ; palm oil for our soap ; kernels which presently will find their oil labelled " fine salad oil," or " rich margarine." The sundries, too, are there by the waggon load ; hemp and cotton, ground-nuts and skins, ebony and ivory, a veritable river of produce flowing into the heart of the British Empire without intermission. Nothing can check that flow, nothing can stop its increase, for it springs to-day from lands overflowing with forest wealth ; lands where natives are inured to the hardships of labour, natives of infinite patience and withal the world's keenest traders. There is but one danger to this increasing flow—race prejudice—which may, unless checked, give birth to actions which will utterly shatter African confidence in the British race.

The critics of the African all agree that he has one good point—" he takes his gruel like a man "—" flog him when he is in the wrong and he won't resent it ; flog him thoroughly whilst you are at it, and he will even thank you for it." If this doctrine should ever firmly possess the minds of those whose duty it is to administer West African colonies, the Governments will be faced with a danger impossible to exaggerate. To make this opinion an article of administrative faith is to provide the white with a salve for every act of injustice which irritating circumstances and climate so constantly generate. In every colony in West Africa there are some few white men who are wholly trusted by the natives, and their homes and hospitality are at their disposal day and night. Naturally these are the experienced men of the coast, or those of repute amongst the natives ; the easy grace with which they move in and out amongst the people at all hours, and in all circumstances, is demonstrative of the confidence they enjoy. Discuss the natives and the problems of administration with such men and the furrowed brow wrinkles still more, and they tell you a change must come soon, or—" Certain white men would be wise to clear." It is for statesmen at home to recognize the danger in time and choose between a day of reform or a day of reckoning.

PART III

I

LABOUR—SUPPLY AND DEMAND

EVERYWHERE in West Africa the cry goes up, " Give us more labour." The British, German, Portuguese and French merchants all declare that if only they could get the labour, they might put a different face on the whole of the problems of production in West Africa. The principal reason for this shortage is unquestionably the fact that West Africa is sparsely populated, but this one fact does not, by any means, explain the situation. In Liberia alone does there appear to be any appreciable quantity of surplus labour, and upon its resources considerable demands are made by other colonies. This surplus obviously arises from the fact that Liberia is completely undeveloped, but if in the near future some energetic power should take charge of that territory, a period would certainly be put to indiscriminate recruiting amongst the native tribes.

It is true that in some territories in West Africa there is an increase in the population, but taking the whole areas into review, the labour force has seriously decreased within recent years. Statistics, though at present little more than estimates, go to prove that in several colonies this falling off is becoming a grave question. Recently the religious denominations in Lagos have been holding "intercessions" with reference to the high rate of mortality. If this intercession should lead the natives from faith to works, we may still hope to see the abandonment of those European customs which are doing untold harm to the physique of the native women and children.

The causes of decrease in the population, generally speaking, are beyond human ken and one can only express opinions which someone else will promptly contradict. For example, almost every traveller wrecks his reputation on that old-time rock of controversy— polygamy. Sir Harry Johnston mentions in one of his books the case of a polygamist with 700 children, but the greatest polygamist I have ever met in Africa possessed 1000 wives, yet he had no children! Argument based upon two such instances, however, is profoundly unsatisfactory, because with so large a company of wives in one case, and children in the other, it is obvious that many other considerations repose beneath the surface. There is one outstanding fact which everyone knows, but few speak about except in whispers ; human nature is pretty positive in West Africa, no matter of what hue the skin, and scientists may argue until eternity upon the relative effects of polygamy and monogamy on the birth rate, but all their deductions are wide of the mark whilst they have so little actual monogamy anywhere in West Africa.

Sleeping sickness has made the most terrible ravages wherever it has established a firm hold on the tribes, but this scourge would seem to be spending its force. Seven years ago Uganda recorded over 8000 deaths from sleeping sickness within twelve months, and the latest

Government report shows that there has been a gradual reduction until in the year 1910 there were only 1546. Happily this encouraging feature is present on the West Coast also. The Congo suffered more than any other colony, due, probably to a large extent, to the systematic oppression under which the population groaned during the Leopoldian régime. Now, however, the absence of the scourge in many of the old districts is quite noticeable. Villages that we knew to be swept by this plague ten years ago are once more flourishing, and in some cases where the birth rate was almost nil the villages are again joyous with the laughter of little children.

The worst sleeping sickness areas remaining in West Africa appeared to me to be the Bangalla region of the Congo and the Portuguese island of Principe. In the latter it has reached such proportions that the whites are leaving the island. The Portuguese still keep a considerable number of slaves on the cocoa farms, all of them either infected or exposed to the disease. As one passes from roça to roça, these slaves, stricken with disease, with emaciated bodies and gaunt features, stare piteously at the passer-by from eyes that seem to stand out from their heads, mutely appealing for the freedom of their distant village homes on the mainland. Looking at the matter from the materialistic standpoint of labour-supply, but makes this ruinous conduct on the part of the Portuguese appear doubly reprehensible.

" Civilization," too, has contributed to a decrease in the working population, but in a varying degree. All the Powers have sinned in this respect. I never read of punitive expeditions with " many natives killed " without inwardly fuming at the folly of the administration which should know how precious from an economic standpoint alone, is the life of every single native. Yet in some places the tribes are hustled, tormented and even butchered in a manner little realized as yet by the European public. Think of the loss of life by violent death in both Belgian and French Congo, and in German West Africa! Think of the countless thousands of bleaching bones scattered over the highways through Portuguese Angola!

Within the last twenty-five years well over 60,000 slaves have been shipped to San Thomé alone; add to this the thousands sold and still in slavery on the mainland, and you probably have a total of over 100,000 slaves passing into the possession of the whites in Portuguese West Africa. That stream of human merchandize involved a wastage of another 100,000 lives, for a Portuguese slave-trader once admitted that if he got half his total gang to the coast, he was lucky, but that generally he could not deliver more than three out of ten!

It is a haunting thought that since the " 85 " scramble for Africa, the civilized Powers who rearranged the map of the African continent, ostensibly in the interests and for the well-being of the natives, have passively allowed the premature destruction of not less than ten millions of people. Now these Powers complain bitterly that they are short of labour and jump at any expedient which presents itself to obtain labour for their hustling developments.

The sins of King Leopold are visiting themselves upon his successors in every part of the Congo basin.

COCOA FARM, BELGIAN CONGO.

WHAT GERMANY LACKS

The prospective gold mines, the cocoa farms, the public departments, all of them are handicapped owing to lack of an adequate labour force. If only the Belgians could restore to life an odd million of the able-bodied men and women done to death under the régime of their late sovereign, what a different outlook their colony would possess !

The Belgians now propose bringing Chinese for the Katanga Mines, but seeing that their former experience of Chinese coolies was not a happy one, and considering other drawbacks, I very much doubt whether they will ultimately launch the experiment of bringing thousands of Chinese across Africa. The original idea of the Belgian Government was that of bringing the coolies into the Congo under a regulation which would secure their repatriation at the termination of the contracts, coupling that regulation with others similar to those adopted by Great Britain in South Africa. Mr. R. C. Hawkin, * whose knowledge of South African politics is not only wide, but intimate, at once pointed out that the Belgian Administration was restricted by the Berlin and Brussels Acts. This opened up a situation so obviously awkward that nothing more has been heard about the introduction of Chinese labour into the Congo, at least for the present.

Germany, like Belgium, differs from France and England in that she has no other colonies from which to draw a labour force. Quite recently her colonists, at their wit's end for labour, passed a resolution agreeing to import 1000 Indian coolies for labour in the mines. It had not occurred to them that the British India Office

* Secretary of the Eighty Club.

might object. How much trouble, to say nothing of expense, they would have saved themselves if only they had asked the office-boy in Downing Street !—they need have gone no higher.

This is another instance of the strange features which now and again attend German colonization, good as well as bad. Their authorities had apparently entirely forgotten the regrettable Wilhelmsthal affair, but probably the real reason was that this incident (which many Englishmen will not readily forget) was regarded by them as altogether too trivial to be noticed. This unfortunate affair—though in some respects comparatively unimportant, yet in reality a grave matter—certainly merits a permanent record in some form, because it is just one of those blundering incidents which bring in their train a whole crop of labour difficulties.

A German Railway Construction Company had been allowed to recruit British Kaffir subjects from South Africa. In the autumn of 1910 trouble arose because deductions were made from the labourers' wages, and they further complained of bad food and housing. The Railway authorities seem to have then embittered the situation by refusing to allow the men food and water. This conduct in a tropical country was little, if at all, short of inhuman, and the labourers naturally struck work and apparently assumed a somewhat threatening attitude. The situation was then handled in a style characteristically German. The Company itself, ignoring the civil authorities, called in the troops, who shot seven of these British subjects in cold blood and wounded several others. How one-sided the whole affair was is demonstrated by the fact that not a single German soldier was even injured.   This incident, from every point of view an outrage, was regarded as so trivial that no one appears to have been punished, nor so far as we know has any compensation been paid to the wounded or to the relatives of the murdered Kaffirs.

German colonial knowledge of British public opinion cannot be of a very far-reaching nature when it ignores this incident in asking for British labour to develop its colonies.   To Englishmen it cannot be a matter of surprise that the India Office has not yet granted permission to recruit labour from the Indian Empire.

Germany and Belgium are the only two Powers in West Africa which do not possess colonies in other parts of the world from which to recruit labour, hence they are dependent upon other Powers.   To the proud German Empire, this situation is irritating, while Great Britain, France, and also Portugal, to a limited extent, can each of them augment the labour force of any given colony by recruiting from their other colonial possessions.

The Portuguese colonies of Angola, San Thomé and Principe, which comprise the major portion of Portuguese West Africa, experience the greatest difficulty in obtaining labour.   It is perfectly true that during the last half-century, close on a hundred thousand labourers have left the shores of Angola for the cocoa islands and other places, but these it must be remembered were almost exclusively slaves which had been bought or captured in the remoter regions of Angola, Rhodesia, Barotseland, and, more especially, the Congo Free State.   The Portuguese colonists of Angola are so pressed for labour that they started some years ago an "anti-slavery" movement against the Portuguese planters of the islands. No doubt there was an honest element in this movement, but it is equally beyond question that the mainspring of the movement was local anxiety to keep all the slaves in the Angola colony, which is to this moment rotten with slavery. If Angola, a territory more than twice the size of France, were properly developed, it would require first of all a complete abolition of slavery, and then an immense augmentation of the labour supply. When we were at Lobito, the Robert Williams Railway Company and the Electrical Syndicate between them were at their wit's end for two thousand more men, but these could not be obtained.

The two colonies of San Thomé and Principe are by far the most serious problem. The area of the two islands is not large—only 400 square miles together—but they are extraordinarily fertile ; the very air seems to intoxicate with abounding fertility ; everything flourishes, cocoa, sisal and rubber ; everything multiplies and replenishes on the earth, but man ; for some reason there appears to be a curse upon those islands, they are almost without an indigenous population and the wretched slaves imported to fill the ranks die off like flies. The future of the Portuguese cocoa colonies is doubtful because it is obvious that they cannot be run permanently by a temporary solution of the labour question.

Both France and England at present manage their labour difficulties with greater ease than any of the other Powers, and this because both have a floating supply in their colonies, which, owing to the high standard of colonial development as expressed in railways and steamers, motors and good roads, is readily transferred to the more needy districts. At the same time every now and then we hear laments that expansion is rendered impossible owing to the lack of men.

When Lord Sanderson's Commission took up the study of contract coolie labour, the areas appealing for labour included the Gold Coast colony, and the Government Secretary of the mines, Mr. Cogill, put in a plea that the colony should be allowed to recruit labour for its mines from India. In this plea he was supported by Sir John Rodger and the Acting-Governor, Major Bryan. This application is not easy to understand, for everyone knows, or should know, that Indian labour generally is unsuited to mining work. There is, however, some reason to believe that the inspiration of this plea came from sources requiring indentured labour from other parts of the world, and that the demand for Indian coolie labour was put forth in the hope of establishing necessity and thereby paving the way for a less acceptable demand.

The bulk of labour in West Africa is employed under indenture or contract, the majority of the latter being for three years, but a great deal of unskilled labour is employed on a yearly, or in some colonies—particularly the Portuguese—a five years' contract. The latter are paper contracts, and in practice may mean anything or nothing at all. Very few unskilled labourers in Africa are prepared to accept willingly a single contract of longer duration than one or at the most two years, and if a contract system exists whereby labourers are bound for longer periods at a single service, it may be generally assumed that some form of pressure or intrigue has been at work.

Now that public attention is being focussed upon labour conditions, it becomes increasingly imperative that Governments should lay down the broad lines upon which they are prepared to allow contract labour. Nor must the labourer only be considered, the employer has the right to be heard in framing such conditions. In spite of much evidence to the contrary, I am still inclined to the belief that, as a class, the employers of labour everywhere in Africa detest as much as anyone labour conditions which are unfair. Even the Portuguese planters of San Thomé hate the slavery they practice, but by a long series of blunders they have been led into their present position.

The greatest care requires to be exercised if contract labour is to be kept free from the taint of slavery. The Indian authorities, in spite of every precaution, frequently find that the most reprehensible· practices attach to the recruitment of labour for the East and West Indies. In the African continent, where domestic slavery is so widely prevalent, the need for watchfulness is a hundredfold greater.

The conditions which govern the immigration of indentured labour should differ but little from those which cover local contracts, with the one exception that local labour contracts should always be of short duration—never longer than a year. Contracts for over-sea labour must be longer to cover the cost of transport, but even these are seldom satisfactory to either employer or employee for a period longer than three years. The Jamaican and Fiji indenture, which in practice involves a contract of ten years, is for many reasons highly objectionable.

The chief danger is beyond question with the recruiters. In India these men according to Mr. Broun,—an Indian Civil servant of large experience—" are the worst kind of men they could possibly have. They are generally very low class men." They seem to bribe, deceive and bully by turns, anything indeed to bring the Indian coolie into their toils. In Portuguese West Africa the recruiter has for years been a slave-trader pure and simple, purchasing slaves from the Congo rebels, and also from the chiefs in the Rhodesian borderland. The Portuguese Government has now issued a regulation that all such recruiters must be duly licensed. In Belgian, German and French colonies, recruiting is undertaken very largely by Government officials.

Recruiting—whether by the irresponsible recruiter, the licensed agent, or by the Government official— calls for the closest attention of the Administration. The official will demand from a chief, the unofficial recruiter will bribe him for a given number of labourers ; in the former case the chief fears to refuse, in the latter he becomes a party to a form of slavery.

The German official carries this operation through with the least amount of sentiment. I asked a planter in the Cameroons whether he obtained all the labour he wanted with a fair amount of ease. He looked at me in astonishment, and replied, " With ease, of course. I only notify the Government that I want labour and they

bring it to me!" On another occasion, when I was
discussing Portuguese administration with a French
cotton planter from the Cunene, he began roundly abusing
the Portuguese Government, and upon my inquiring
wherein they differed from the German administration
across the river, he replied, "The Germans stand no
nonsense over labour. If the native villages are small
and distant from the planters, they just burn down the
villages and drive the natives nearer the planters. The
Government can then quite easily make a list of the
able-bodied men and supply them as they are required."
How far this may be a general characteristic of German
treatment of native races, I cannot say, but what I have
seen of German colonial methods does not impress me that
their occupation is far removed from a sort of military
despotism. In the matter of official recruitment of
labour, the Germans are by far the most vigorous of any
of the West African Powers. In this official recruitment
the individual labourer concerned has very little say
indeed; that he should desire to enjoy his freedom is
apparently no concern of anyone, all he knows is that
he has to work for the white man for a given period,
and in German South West Africa the "contract" must
be made "as long as possible."

The hardships of contract labour are greatly increased
by the prevalence of domestic slavery. We are sometimes told that domestic slavery is inseparable from
native social life, and that from time immemorial it has
been an integral part of African law and custom. For
that matter so has cannibalism! There are many
apologists for domestic slavery, including students of

such eminence as Mr. R. E. Dennett and the Editor of the *African Mail ;* the latter considers it would be foolish to abolish the House Rule Ordinance—or in other words the legalization of domestic slavery in Southern Nigeria. It is difficult to understand how a man with Mr. Dennett's experience could possibly write the paper on this question which was reprinted in a journal of the Royal Colonial Institute. Mr. Dennett knows, or should know, that the horrors of early history in the middle Congo, the blood-curdling stories of Kumasi, the present-day slavery of the Portuguese colonies and a thousand other labour scandals rested and still rest in the ultimate resort upon domestic slavery. The cheap sneers at the sentimentalist, the innuendo that they are mere stay-at-home critics is entirely misplaced and no one knows this better than Mr. R. E. Dennett.

Domestic slavery is slavery pure and simple, although I agree that under the African chiefs it may not be so bad as under the old planter systems. Front rank statesmen with large administrative experience have recorded the lamentable results attaching to domestic slavery, and so recently as 1906 Africa's greatest constructive Administrator—the Earl of Cromer—penned the following significant passage :—

" If the utility of the Soudan, considered on its
" own productive and economic merits, is not already
" proved to the satisfaction of the world—if it is
" not already clear that the reoccupation of the
" country has inflicted, more perhaps than any other
" event of modern times, a deadly blow to the
" abominable traffic in slaves, and to the institution

" of domestic slavery, *which is only one degree less hate-*
"*ful than that traffic*—it may confidently be asserted
" that we are on the threshold of convincing proof."*

The broad lines of domestic slavery are common
throughout West Central Africa. The slave becomes the
property of the head of the house or chief, who can
" contract " him to third parties without reference to
the one primarily concerned, that is to say the slave
himself, who in turn cannot hire out his labour without
the consent of his master, and he may also be transferred
in payment of debt. Upon the death of the owner, the
slaves with their families—who are the property of the
chief—are divided amongst the heirs with other goods
and chattels of the deceased. The domestic slave can
by native law everywhere, and by European law in some
parts, be recaptured if he runs away. According to
British law the slave becomes the property of the master
in Southern Nigeria " by birth or in any other manner."
This only legalizes native law, it is true, but " in any
other manner " throws the door widely open to a transfer
of human beings in a way highly repugnant to British
sentiment.

In the middle Congo a system is rapidly extending
which violates every moral code in that it is none other
than a wholesale prostitution. Under this custom,
known locally as that of the " Basamba," a man hires
out a proportion of his wives on a monthly or yearly
agreement. The basis is the principle of absolute ownership ; a weekly or a monthly " hire " in cash, or its
equivalent, is paid, and all the offspring handed over

* Italics mine.—J. H. H.

A CONGO CHIEF WITH SOME OF HIS WIVES AND "BASAMBA" CONCUBINES.

to the husband and owner. Thus the owner, or husband, obtains first a financial return for the hire of his surplus wives, and secondly he claims the offspring. In the event of males, they become domestic slaves, with which the chief may satisfy administrative and other demands for labour ; while in the case of girls the chief possesses a further source of revenue either by hiring them out to "temporary husbands," or by purchasing other and older women for the same purpose. This method of increasing the number of wives and slaves is by no means limited to the middle Congo, but in no other part of West Africa were we able to find it carried on so extensively. In those regions it is quite common to find men with ten wives hired out in the different villages, and a few cases exist of men who now carry on their trade with no less than fifty, and even one hundred, such surplus wives !

There are other means of obtaining domestic slaves. Many of them are, of course, "inherited," and not a few are passed over as part dowry with a wife ; others are taken for debt and some are captured in tribal warfare.

The relation between contract labour and domestic slavery is more intimate than appears on the surface. In West African practice an employer desiring a given number of labourers invites or "calls" the chief whom he informs of his requirements ; if a merchant, he generally accompanies his request for boys with a gift ; if a Government official, the demand more often than not is accompanied by a threat. At a later date the chief returns with the required number of labourers. If asked whether they are willing to work, they generally assent, for they fear to oppose their chief who, even if European prestige were not behind him, still possesses all the power of native law and customs—to say nothing of the awe-inspiring fetish.

Admittedly, however, normal domestic slavery in Africa is widely removed from predial slavery with which our school books made us familiar. Eliminating from domestic slavery the sacrifices for which slaves were always, and in some places are still, reserved ; eliminating also European demands for labour, the system is not everything that is bad, nor are the chiefs invariably cruel and despotic towards their slaves. It is nevertheless equally true that the frequency of " palavers " which deal with escaping slaves is an evidence that the yoke of slavery is often intolerable, and that in spite of native law, in spite of European law and practice, and still more in spite of the fetish, the slaves attempt, and sometimes make good their escape.

Over large areas in the British colony of Southern Nigeria the police can, and do, recapture and restore such slaves to their owners, and two years ago it came as a shock to many that an escaping slave seeking refuge on the deck of a British Government ship could be forcibly recaptured and restored to his master ; not only so, but he was actually flogged by British police for running away ! It is, however, not altogether an easy matter to secure recapture of runaway slaves under British law, and therefore to the charge of " running away " is sometimes added larceny—the theft of a canoe or a cloth ; the canoe, of course, being the boat by which the wretched slave made good his escape, and the cloth that which he

A HUNTER'S "LUCKY" FETISH.

uses to cover his nakedness. The following is a fair specimen of the warrants issued for the recapture of slaves in Southern Nigeria :—

COPY. No. 1881
—————
74

WARRANT TO ARREST ACCUSED.
Form 2.

In the Native Council of Warri, Southern Nigeria.
To..............................Officer of Court.

Whereas Joe of Lagos is accused of the offence of (1) running away from the Head of his House two years ago ; (2) Larceny of cloth value 16s., two handkerchiefs, and a canoe. You are hereby commanded to arrest the said Joe of Lagos and to bring him before this Court to answer the said charge.

Issued at Warri, the 28th day of November, 1910.

(Signed) : PERCY GORDON,
Senior Member of Court.

The British Government alone amongst the Powers in West Africa really dislikes this system and shows some inclination to secure its abolition. The Portuguese like it, and in the main descend to the level of it, manipulating the system to suit, so far as possible, their labour requirements. The Belgians cannot recognize it without violating the Berlin and Brussels Acts, so they leave it alone to bring forth a whole crop of abuses.

The Lieutenant-Governor of French Guinea has recently taken a strong line upon the question of domestic slavery, which other Governments might emulate. He has issued instructions to all his subordinate officials in which he says :—

"We cannot allow the system of captivity to "continue any longer ; it is a matter of duty as "well as of dignity to put an end to the present "situation. . . . You are to profit by every occasion "which offers for making the captives understand "that it is immoral for one man to possess another. ". . . Whenever you or your colleagues make a "journey you are to gather the natives together "and explain to them our wish. . . . In all cases "which are brought before you, you are resolutely "to refuse to examine those which relate to master "and slave ; make them understand that for us "there are no slaves, and that in justice and law we "only admit the relations of employer and employee. "You are to follow up with the utmost rigour all "crimes committed against human liberty, and to "employ all the severity of the laws against barbarous "masters or slave-traders who are still too numerous "on the frontiers of neighbouring colonies. . . . "Every captive who appeals to your authority is "to be welcomed by you and protected against "every abuse of force. You will disregard every "stipulation which in civil contracts, wills, etc., "would postulate the condition of family captivity. ". . . There are no longer any captives in Guinea— "such is the formula which must rule your conduct."

If transfer to French Congo is a promotion, the

quicker the French Government promotes this enlightened official to that sphere, the better for French reputation in that unhappy region.

In Africa forced labour, like contract labour, rests very largely upon domestic slavery. What is generally understood by forced labour is indistinguishable from the corvée of Germany, or from that which obtained in earlier times in Prussia and France. It is simply a communal undertaking upon works of general welfare, mainly roads from town to town, although the word corvée was also applied to all feudal demands, but in those cases some wages were given in return for the labour.

The old African communities exacted, and in many cases still exact, labour from their domestic and agricultural slaves for which they were and are paid, according to the whim or the benevolence of the chief. This labour was, and is, devoted to the clearing of paths, keeping bridges in repair, gathering harvests, porterage, canoeing, boat-building, and indeed any undertaking which involves a considerable labour force. These exactions, however, are always made at a time which avoids interference with agricultural necessities ; moreover, in the nature of the case, the labour was never used very far from the village.

European administrations have stepped into West Africa, and have taken the place of the chiefs, and in so doing have adopted corvée under the plea of works of public utility—a blessed phrase which covers a multitude of questionable " necessities."

In the Gambia every able-bodied male is compelled under the penalty of a fine, or six months' imprisonment, to give labour for the construction of roads, bridges, wells and clearings round the villages in his own district.   They must also provide carriers when required.   Apparently the Governor is the only arbiter of the time to be given to such works and whether or not any remuneration may be made.   In Southern Nigeria the Governor may call up all able-bodied males between 15 and 50, and all able-bodied women between 15 and 40, to give labour upon road-making and creek-clearing for a period of six days each quarter.   Refusal to obey involves a fine of £1 or imprisonment not exceeding one month.   Similar regulations prevail in Northern Nigeria.

In German Togoland the natives must give twelve days a year, or commute this by paying six marks ; but the labour can only be used upon roads and bridges in the district in which the labourers reside.   Almost identical regulations prevail in the French and Portuguese Congo. These regulations—*qua* regulations—are unobjectionable and, after all, only assume powers exercised for generations by the chiefs.   In practice, however, under the term works of " public utility," frequent and irregular demands are constantly being made to the irritation of the people. Think of what a single punitive expedition involves— no matter on how small a scale.   Modern weapons of warfare, ammunition, tent kits, provisions and the thousand and one odds and ends of the modern paraphernalia of war, all this is carried in the main by forced labour.   I shall doubtless be reminded that the chiefs always exacted labour for war.   That I admit, but " civilized warfare " is so infinitely more elaborate than the simple native spear and arrow warfare, that they are not to be put in the same category.

Carriers too are demanded in numbers and for distances which violate every native restriction. It is but two years ago that a British official in Southern Nigeria decided to start off upon a journey on Sunday morning, and because the carriers did not come quickly enough, he marched into the two nearest churches and seized the congregations, including the native minister, and to demonstrate further his petty authority and repugnance of loftier ideals, insisted on this native clergyman carrying a box containing his whisky. At this distance it is the ludicrous which probably strikes the imagination, but it is an entirely different matter locally. The missionaries of Southern Nigeria, no matter what their denomination, are of a very devout and noble-minded order ; they have instilled into the minds of the natives a deep reverence of all things pertaining to worship, and nothing will ever efface from the native mind that—to say the least—irreverent conduct of the representative of the Christian Government of Great Britain.

It is difficult sometimes to discriminate between contract labour, forced labour and slavery, the boundary lines having been obliterated by vigorous administrations demanding labour for this and that work of public utility, which in reality bear little relation to an enterprise for the general welfare. In Belgian Congo this is carried further than in any other West African colony. The Belgians insist that there is no forced labour in the Congo, and this is perfectly true from the legal point of view, but nevertheless almost the whole administrative machinery and Government undertakings are maintained by forced labour. To roads and bridges Belgium has added telegraphs, mines, plantations, and recruitment for the army ; the ranks of both—labourers and soldiers—being filled almost entirely by forced labour.

Loud were the complaints made to us in our recent journeys through the Congo of the incessant demands for labour by the Administration.

Wearied with a day of struggle through Congo forests and swamp, I was resting one moonlight evening in the centre of a primitive Congo village ; a group of native chiefs were sitting round me discussing political conditions. The absence of a certain token led me to question one individual somewhat pointedly as to the cause.

" If I tell you, white man, you won't betray me ? "

" Your chief knows me well enough for that," I replied.

" Well, there were eight of us," he explained, "called by Bula Matadi. We were bound to go, bound to leave our wives and our children and go down river several days by steamer. When we arrived, the head white official gave us a ' book ' (contract) for three years, and sent us to cut a road for the ' Nsinga ' (telegraph wire). We worked for some days, discussing every night how we could escape. One afternoon the white man · went into the forest and four of us who were working together ran down to the river where we found an old canoe and one paddle hidden in the grass. We crowded in and pushed off, one guiding the canoe with the single paddle, whilst the others paddled with their hands. We managed to get into a creek hiding ourselves until the next night, when, with the help of some stout sticks for paddles, we began the long journey home, paddling in the night and hiding ourselves and our canoe during the day. We lived on roots and nuts for eight days, and then, when hiding in the forest, we heard some women talking we ' frightened ' them and they fled, leaving their baskets behind. These contained palm nuts, on which we lived for another six days. On the fifteenth day we reached home again, but our people did not at first recognize us."

" Why ? "

" Because, white man," chimed in the old chiefs, " they were so emaciated that the flesh had shrunk from their cheek-bones, their ribs stood out like skeletons, and they could barely speak."

Such is Belgian forced " Contract Labour " in the Congo.

What are the boundary lines between legitimate forced labour and that which public opinion, as trustee for native rights, should refuse to tolerate ?

The broad line of division is unquestionably between genuine works of public utility on the one hand, and profit-bearing works on the other.

Road-making, bridge-building, creek-clearing, are all of them works from which the whole community benefits, but the requisition of this labour should not be left to the arbitrary will of a temporary official, but subject to clearly-defined regulations. Any legislation upon forced labour for works undertaken for the public good should only permit the requisition, in lieu of taxation, as is the case in German Togoland, where the native has the alternative of paying a head tax of six marks per annum or giving his labour for twelve days, subject to the labour being required for the improvement of his own district.

As in Ceylon and other British colonies, the natives should be allowed to commute the labour by a money payment. To labour exacted under these rigid conditions, there can assuredly be no strong objection, and, generally speaking, the native tribes would loyally co-operate in such proposals.

To employ forced labour upon any kind of work which carries with it a financial advantage partakes of slavery. A merchant obtaining forced labour at his own price is thereby, in principle, engaging in slavery, and if by obtaining such labour he is able to enter into unfair competition, he is further guilty of doing a gross injustice to his fellow-merchants. The Belgians are extremely prone to this form of labour. In the Congo there is a good deal of State commercial enterprise, which may yet ruin the individual merchant. The Belgian Government is doing the larger proportion of transport on the vast fluvial system of the Congo, and thereby competes with the Dutch House and other transport companies.

These transport steamers are all driven with wood fuel cut from the forests. Every few miles along the banks of the Congo river there may be seen stacks of fire logs cut into lengths of about eighteen inches, which have been either cut by the employees of the Government or by the villagers. No company is permitted to purchase Government wood, and ordinary steamers purchasing from the villagers have to pay 2 francs a fathom for such fuel. Journeying down the Congo a few months ago, three of us carefully examined conditions at one of the wooding posts, manned by twenty-six men and ten women, most of whom had been " demanded " from the

chiefs in more distant parts of the Congo, and drafted to
the spot in question.   Several had already served three
years—the nominal term of the contract—but, without
any option in the matter, their contracts had been
renewed.   Each of the men had to cut one fathom of
wood per diem ; some were paid 7 francs and others only
5 francs a month, with a 3-francs allowance for food.   The
maximum cost, therefore, was 10 francs for the thirty
fathoms of wood cut in the month.   Thus the State
provides itself with wood at a fraction over threepence
per fathom, for which company steamers must pay
2 francs.   Under such systems not only are human
liberties violated, but commerce suffers prejudice.   There
is not a little danger that the Belgian authorities intend
giving a considerable extension to State enterprises, which
in all probability will be prosecuted with this form of
forced labour.

The question of State railways and telegraph lines is
a difficult one, both partaking of works of public utility,
yet both are as a rule profit-bearing.   There is the further
consideration that all profits go to relieve local taxation.
Given representative Government or given even an elective
element in the Administration, there may be some justice
in imposing this form of forced labour upon the general
community, but under the autocratic systems of Crown
Colony Administration, large demands for forced labour
cause, not unnaturally, widespread disaffection.   Fortunately British colonies are almost entirely free from the
employment of such labour and to this no doubt is due
the excellent management of all railway systems under
British control.

The most economic and the most politic line to follow is that of the employment of free labour. Supervision is reduced to a minimum, abuses of authority are rare, the work goes more smoothly, the song takes the place of the boot and the lash, the native labourer goes home when the day's toil is over vowing vengeance on no one, and the white man returns to his somewhat primitive home with a mind undisturbed by conscious wrong-doing.

II

LAND AND ITS RELATION TO LABOUR

It will not, I think, be contested that throughout West
Africa there is no native conception of private ownership
of land. This is almost an article of religious faith
amongst the African races generally. Let one tribe
murder a member of another community and a palaver
will be called and compensation paid. If wife-stealing
or kidnapping of boys takes place, the tribes involved will
remain calm and settle their dispute by making peaceful
and honourable amends. Let one tribe exploit the palm,
or without leave settle on the lands of another, and, on
the instant, the ultimatum is despatched—" Depart
forthwith, or accept the alternative ! " Indeed the occupation of the communal lands of another tribe is recognized by most tribes as an overt act of warfare, the signal
that all negotiations for peace are at an end.

Perhaps no more eloquent testimony of the attach-
ment of native tribes to their lands is to be found anywhere than in the great Equatorial regions of the Congo.
The early 'eighties witnessed in the Congo basin three
convulsive movements ; the entrance of the white man
from the west, following on Stanley's journey across the
continent ; the incursion of the Arabs from the north,
and the Lokele wars towards the south. This latter

movement was destined to change the whole situation in the Equatorial regions, south of the main Congo. The Lokeles, probably pressed by the Arabs from the north, started a " land war " with their southern neighbours, the object being to obtain an extension of tribal land. This pressure set in motion a land war, which ultimately extended over an area nearly five times the size of Great Britain and ran right through the south reaching down to the Lukenya river, and in some places even across the greatest of the southern tributaries— the Kasai. Tribes fought each other for the maintenance of their ancient boundaries until the whole of the Equatorial region was in a state of warfare, which only ceased when starvation claimed victims by the thousand. Then only were boundaries re-adjusted by peaceful agreements; even so the whole population for months was in such dire straits for food, that men sold their wives, and mothers their children, for a single basket of manioca. One realizes how passionately the natives are attached to their lands as they recount the horrors of those terrible years. Said one to me recently—" At first we fought to protect our lands, but in the end we had to fight to obtain ' meat '—human flesh—to stay the pangs of hunger."

The native boundaries are almost invisible to the European eye, but to the African student of nature those boundaries are fixed and immovable as the eternal hills. The limits of tribal lands, within the orbit of which the clans may move and hunt whenever they will, are the stream, the palm plantation, the hilly range and the bridges across streams and rivers.

Upon the chief and his advisers devolves the sacred duty of maintaining intact these tribal lands, alienation being foreign to the native ideas. So jealously is this guarded that many paramount chiefs in native law have no power to grant even occupancy rights. For six months the cession of Lagos to the British Crown was held up because King Docemo had signed a treaty which appeared to violate this principle of native law. The population declared that the ownership of the land of Lagos was not vested in the paramount chief, but in the seven White Cap chiefs, who, fearing the terrible consequences of alienating the tribal lands, fled to the bush. It became necessary for the British representatives to give the most explicit assurances and sacred promises on the point, in order to secure the ratification of the treaty of cession.

It is perfectly true that titles have been granted to native tribes and to white men, but it is equally true that originally there was never the remotest idea that this involved the European conception of total alienation. In the Holt v. Rex case of Southern Nigeria, the Crown held that "under native law strangers cannot obtain freehold rights—only occupancy rights." The tribal conception of occupancy rights also carries with it the communal idea; a native clan settling by permission within the territory of another tribe really constitutes the first step in progressive incorporation. In the first instance of white settlers, there are abundant stories of the native interpretation of this principle—some of them distinctly objectionable, although there were pure motives behind them; others are amusing, such as that of the chiefs "borrowing" saws, axes, string, rope, nails and what not. Again and again they have freely and openly helped themselves to palm nuts and other produce from the white man's ground. No doubt much of what the European calls "pilfering" was really quite innocently founded upon the communal conception of the primitive races.

The impetuous scramble for African territory, which began thirty years ago, made, and continues to make, a considerable breach in this old primitive system. White men, acting through the doubtful medium of interpreters not infrequently corrupted in advance, have secured from chiefs titles to land of all dimensions. These chieftains, as a whole, never fully grasped the meaning of the titles obtained with honeyed words, and which they are now unable to repudiate. That this is so is partly proved by the fact that in some colonies areas have been conceded twice and even three times over. Swaziland is, of course, the most flagrant example, where it will be remembered a situation so complex was created that it ultimately became impossible for any Court to decide as to who were the real owners of specific areas.

In West Africa things are not, and never can be, quite so bad, although in some colonies, the Gold Coast for example, German Cameroons and French Congo, land difficulties are being piled up for the endless confusion of future administrators. In Belgian Congo there is no immediate probability of trouble, due partly to the fact that capital has little confidence in Belgium's heritage, but more because the major part of the population has disappeared.

There is a vital connection between land and labour

in all tropical and sub-tropical colonies. The economic future of native races is immobilized in the proportion in which their lands are taken from them. The almost phenomenal success of the cocoa industry in the British colony of the Gold Coast is due entirely to the fact that the natives are the proprietors of the cocoa farms. Throughout the colonial world, there is no more striking contrast between a landed and a landless native community than the British Gold Coast colony and the neighbouring Portuguese colony of San Thomé. In both territories cocoa flourishes, both produce excellent cocoa, in both nature is very kind, but while the one will march on conquering the cocoa markets of the world, the other is doomed to ultimate disaster.

The San Thomé cocoa producer is only a labourer—in fact a slave—and he is perishing at such a rate that the depleted ranks must be filled from outside sources to the number of 3000 to 4000 labourers every year. This constant inflow of labour cannot continue indefinitely, even if European sentiment permitted—which it will not—the revolting concomitants by which this labour has been maintained. The economic future of these colonies from which the supplies are drawn will soon forbid the emigration which at present is necessary to the island of San Thomé. The population of the Gold Coast, on the other hand, happy in the enjoyment, in the main, of its own lands, reproduces and to some extent even increases itself every year. The native occupies his rightful place as producer, while the white man finds his true sphere, first as the inspirer of native efforts to place on the market cocoa of increasingly good quality, secondly as the medium by which the cocoa produced is conveyed to the manufacturer, and thirdly that by which surplus European manufactures are brought to the door of the native in exchange for his products.

This relationship of land to labour is receiving increasing recognition by students of colonial policy. The Republican Government of Portugal, finding both labour and land problems in hopeless confusion in the African colonies, has recently introduced a comprehensive measure embracing both factors in the development of African colonies. The ordinance is probably too generous in proportions to be carried through effectively in any colony, and stands little chance of complete application in Portuguese colonies, which suffer already from an excess of legislation, coupled with a rooted contempt for " Lisbon dictation." This new ordinance, however, is a valuable contribution to West African legal literature.

The Provisional Government first lays down the proposition that every native in the Portuguese colonies is under " a moral and legal obligation to work." The proposition upon land is in the following terms : " In all the Provinces beyond the seas, wherever there are public lands vacant, uncultivated, and not used for any special purpose, natives may occupy and cultivate them subject to conditions laid down in the present ordinance."

The native in Portuguese colonies, therefore, must work. The sphere of labour he may choose, but idleness is henceforth a punishable offence.

Women, sick men, minors under fourteen years of age, chiefs and those in regular employment, are either exempt from the operation of the ordinance or deemed to have fulfilled its obligations.

Any native may contract his services, but, in the first instance, for a period limited to two years. The agreement is null and void unless the wages are fixed and recorded in the contract. Any clause giving the employer the right to administer corporal punishment likewise renders the contract invalid. The engagement may be made with or without the assistance of Government officials, but any document signed in the presence of a Government authority carries with it both the right and the responsibility of official intervention in any subsequent dispute between the parties. If, however, the contracting parties enter into the agreement without reference to the authorities, the employer cannot look for official assistance in disputes with the employés, although the latter under all circumstances may rely upon official protection and assistance. All contracts must bear the impress of the labourer's thumb. Wages may not be withheld, nor may pressure be exerted to force merchandize upon the employé in lieu of wages.

Recruiting agents must obtain a licence from the Governor of the province, and any infraction of this section of the ordinance is punishable by a fine of £100 to £1000. A heavier penalty still awaits any recruiting agent who attempts to contract labourers for prescribed regions : presumably that death-trap of Portuguese colonies, the island of Principe. The punishment for such violation may be imprisonment for one year, a fine of £200, and at the expiration of the term of imprisonment, expulsion from the colony. Similar penalties await any agent contracting labourers beyond the bounds of his judicial area.

The Republican Government evidently realizes that contract labour, however benevolent it may be made to appear on paper, is not always a heavenly condition, and that the labourer may repent of his bargain before expiration. Section 18 provides for almost every concomitant which attaches to restrained labour. The pill, however, is sugared by a preliminary and somewhat unctuous preamble, that the whole trend of employment must be that of " moral education." In pursuance of this laudable object, powers of arrest are conferred, " precautions " against running away are permitted, and if a second offence occurs, the offender, " when caught," may be taken to the authorities " to be chastised." There are, however, certain limits to these powers, for the employer may neither shackle nor chain an employé, nor may he deprive the labourers of food, nor impose any fines which involve deductions from wages.

If the native of the Portuguese colonies dislikes the yoke of any master, he may, like Adam, " till the soil," for, as already stated, all vacant public and uncultivated lands are at the disposal of the colonists. The first general restriction is that this liberty is only open to those " who do not possess immovable property to the value of £10." The object of this restriction is nowhere elucidated, but apparently it is that of fixing the population upon definite areas.

If, then, the native does not possess immovable property to that amount, he may occupy a piece of land measuring $2\frac{1}{4}$ acres for himself, and an additional acre for every member of his family with the exception of males above fourteen years of age.

A man with two wives, a mother, three daughters, and also three sons under fourteen years of age, could occupy under this regulation a little over ten acres, but the occupation must be an effective one. A dwelling-house must be erected, and two-thirds of the area must be under cultivation, otherwise the title becomes void, and the authorities will expel the occupants. The right of occupancy is inalienable.

During the first five years of occupation, the colonist is exempt from all dues, but at the close of this period taxation is levied and may be paid either in cash or kind. Failure to pay these dues renders the occupier liable to eviction without any compensation for improvements.

After an occupation of twenty years, characterized by the fulfilment of all legal responsibilities, the occupier automatically acquires the freehold. These cultivators or small holders are exempt from serving either in the army or the police ; they are likewise freed from any form of forced labour, hammock carrying, or paddling, but they are not exempt from taking part in military operations with their respective chiefs, when such expeditions are undertaken by command of the authorities.

District commissioners, civil and military officials are urged to induce natives to avail themselves of the land provisions, and are empowered to assign them plots of land. They are also instructed to prepare local regulations safeguarding the rights of the colonists, compile land registers, etc., for which no fees are to be exacted from the natives.

If a native will not labour for another, if he will not sow a field or trade in produce, if in short he is only prepared to stretch forth an unwashed hand and mutter " Matabeesh, Senhor ! " then the official representative of the Government will deal with him. The danger is that other than " wastrels " may be swept into the official net, particularly whilst such operations are so highly profitable to the Portuguese colonies.

First the delinquent is summoned to answer the charge of idling without visible means of support ; then the paternal authorities are to read him a homily on " moral education," and forthwith despatch him to a place where work is waiting for him. If he still refuses to work he may be sent to " correctional labour." There he will receive food and lodging and be given one-third the market rate of wages. " Correctional labourers " may, according to Section 58, be hired out by private persons upon the same terms as the prisoners of State. Such persons willing to employ " correctional labourers " are requested to make formal application, but only those are eligible to receive such labourers who have never been convicted in any court. If they receive such labourers a given sum *per capita* must be paid to the State and a fine of £20 paid for any shortage in " returns " alive or dead, the number hired out must be returned to the Authorities. If, however, escape is feared, the correctional labourers may be returned to State prisons each night.

If the whole ordinance is to be applied to the Portuguese colonies in a measure of completeness hitherto foreign to the Portuguese possessions, then there is some hope that even the leopard may be able to change his spots.

There is little likelihood that the Portuguese land laws will be rendered effective on the spot, especially when we remember that many thousands of miles throughout which such laws are intended to operate are not yet under any sort of administrative control. The step which is finding most favour in British West African colonies is that of declaring all lands, whether occupied or not, as native land under some sort of ultimate trusteeship of the Governor for the benefit of the natives. No purpose can be served by denying that this would place very large powers in the hands of a single individual, even though the powers so conferred may only be exercised " in accordance with native law and custom." It would beyond question give to the Governor powers which in the hands of some individuals might be exceedingly dangerous.

The majority of British Governors of Crown colonies could undoubtedly be allowed to supersede the paramount chiefs in every respect, providing the constitution of the Crown colonies permitted the bringing into full play of this one vital condition, viz. that his actions would always be " in accordance with native law and custom," but Crown Colony government excludes at present any form of representative government which is the un-written law of every African tribe.

Docemo, and his successor Prince Eleko, in Southern Nigeria, exacted, and exact to-day, an abject obeisance from their counsellors, which, if demanded by a British Governor, would secure his prompt recall. No chieftain, whether he be Mohammedan or Pagan, ever enters the presence of the native Council Chamber of Lagos without prostrating himself flat upon the ground and kissing it three times before receiving permission to sit down. Yet this paramount chief could not alienate a square yard of land without the sanction of his advisers.

No British Governor is at present in this position. In practice, his powers under Crown Colony government are in the ultimate resort absolute and uncontrolled, except by question, answer and debate, in the British House of Commons. When, however, the subject-matter reaches this stage, the man on the spot has probably already committed the Government, and the department is therefore bound to defend him.

Admittedly, somebody must protect the native from the wiles of unscrupulous white speculators, no less than from the subtle and treacherous conduct of individual natives. It is the duty of the Governor, as the responsible authority of the Crown and trustee of native welfare, to do this ; let him by all means have power to prevent the alienation of land and to grant occupancy rights, but under a system of government which will give the natives themselves that which they possess by native law and custom—a collective voice in such decisions. It should not be beyond the wit of man to frame a system of governmental control over native tribal lands which would satisfy the great mass of the people, for let it never be forgotten that Africans in the aggregate are reasonable and by no means difficult to deal with along lines which are demonstrably equitable.

PRINCE ELEKO AND COUNCIL, SOUTHERN NIGERIA.

III

PORTUGUESE SLAVERY

In Portuguese West Africa one sees the best and the worst treatment of native races. The best for the free native, the best for the educated coloured man and the best for the coloured woman. In every other colony— and in this respect British colonies are becoming the worst—race prejudice not only prevails but is on the increase. In the Portuguese colonies there is a pleasing absence of race prejudice ; natives of equal social status are as freely admitted to Portuguese institutions as white men ; the hotels, the railways, the parks and roads possess no colour-bar, and if the Portuguese colonies could be purged of their foul blot of slavery, the natives of other African colonies might well envy their fellows in Portuguese Africa. Alongside intimate social relations with the native is a widespread plantation slavery in Angola, San Thomé and Principe.

Angola, one of the largest political divisions of West Africa, is bounded on the north by the Congo, the east by Rhodesia and on the south by German Damaraland ; a considerable section of the northern territory, including the whole Lunda country, comes within the operations of the General Act of Berlin. Apart from the Lunda province and strips of land bordering the rivers, the

colony cannot be said to give any promise of an agricultural future, although if one nation is adept over all
others in turning wastes into gardens, that nation is the
Portuguese, to whom gardens and plantations are second
nature. A Portuguese house without its shady vinery,
its delicate fernery and luxuriant kitchen garden is unthinkable ; even the little children in the streets, instead
of building castles and grottos, find infinite delight in
laying out miniature gardens, in which they arrange
flowers and ferns with artistic taste.

Economically, however, Angola does not pay, its
finances are like many of its old houses—very unstable
and subject to leakage. Walk its streets, visit its
families, Government departments or merchants' houses,
and certain it is that every other man you meet will remind you forcibly of Micawber. The Portuguese community in any part of Angola can be roughly classed as
the Moneylenders and Borrowers. Each, however, appears to be supremely happy and lives in absolute assurance that something will turn up every day to render life
more agreeable.

Loanda, the capital, is a strange admixture of ancient
and modern dwellings, old churches, a roofless theatre
and dilapidated bull-rings. But despite its shortcomings,
the Portuguese have made Loanda the most restful
health-restoring sea-port in West Africa. Boma, the
capital of the Congo, is distant only twenty-four hours'
steam, but it is surely the most unhealthy and the most
foodless place in Africa. The Belgians, if they liked,
might supply fresh provisions to its starving and dying
population,—for everyone in Boma is dying, it is only a

LAND FORMATION, LOANDA, PORTUGUESE ANGOLA.

question of time. In Boma, fowls, eggs, fruit, fish and vegetables are priceless, while every day shiploads can be purchased very cheaply in Loanda, and if shipped twice a week to the Lower Congo, would at least make life, though short, more comfortable.

There is one place every visitor to Loanda should inspect—the old Dutch Church dedicated to " The Lady of our Salvation." Some American dollars would be well spent in preserving this relic, for it is one of the many instances which demonstrate that slaving was a pious occupation in the early seventeenth century. The whole of the interior was once composed of blue and white tiles of pictorial design, and one on the north wall of the chancel is still complete ; this apparently represents the conquest of Angola by the Dutch, who are seen in broad-brimmed hats, braided coat-tails and parade boots, fighting and slaughtering the hosts of savages. The whole operation against the unfortunate infidels is being directed, and presumably blessed, by the Lady of our Salvation enthroned in the clouds.

If Portuguese enterprise has made Loanda a restful spot for weary travellers, British capital—in the person Robert Williams—has turned an unknown strip of desert land into a flourishing sea-port now known as Lobito Bay. It is from this port, with excellent anchorage and transport facilities, that the West Coast will connect with the Cape railway. This Lobito—Katanga railway, though it has only completed some 450 of the 1200 miles to Katanga, promises commercial success when opened, for it should then constitute the cheapest transport route to Rhodesia and the Congo ; that is unless the Portuguese, with their usual short-sighted economic policy, kill the enterprise with tariffs before it has had a real chance of life.

There are only two other ports of any consequence in Portuguese Angola—Mossamedes and Benguella; the latter a harbour with perpetual "rollers" which make a stay on board anything but a comfortable experience. The town itself, like most Portuguese institutions, is going to ruin: the only redeeming feature being the maintenance of its public gardens, fountains and Eucalyptus avenues. Catumbella, an inland town, lies midway between Lobito Bay and Benguella, and with the latter town, constituted the principal centre of the slave-trade. The old slave-compounds and prison-houses confront the traveller in every part of Catumbella and Benguella, and although many have fallen into disuse, some still have the appearance of occasional occupation.

Loanda, Lobito and Benguella all possess "hotels." Those of the capital proper are a strange mixture of cleanliness, tobacco-ash and half-hidden dirt, but at least they are free from the presence of those unfortunate white women who intrude themselves with such persistence on the attention or inattention of passing white travellers in Benguella, and live by running accounts paid irregularly by white men in that most loathsome of all towns in West Africa. Those wishing to visit Benguella should order their rooms months ahead and not be surprised if on arrival Senhor has forgotten all about the order and has neither room nor bed at his disposal. A sound and vigorous rating, however, will generally extort a promise of a room somewhere, a promise which will seldom be

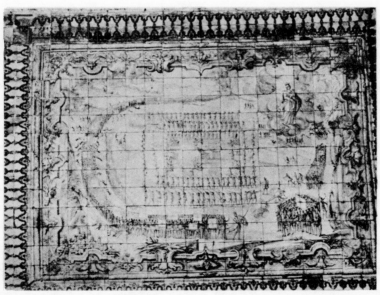

CHANCEL AND NORTH WALL OF DISUSED DUTCH CHURCH, LOANDA.

See p. 171.

fulfilled until all other guests have retired to beds severally robbed of one portion or another to make up an incomplete set for the newly-arrived guests. Nor must the tired travellers be surprised if a black boy enters the bedroom, without knocking, and demands the " other master's pillow," only to be followed later by another woolly pate thrust round the doorway sleepily requesting the surrender of a counterpane or towel, for yet " another master."

It is useless to expostulate with the hotel manager, who will reply with a veritable flood of apologies and threaten to break the head, and neck if necessary, of every black boy in the place, and yet the guest knows with mathematical certainty that he will again have to go through the same course of torture before getting a troubled sleep on that straw mattress in yonder white-washed room. This is the whole trouble with the Portuguese, commercially and diplomatically ; their eternal protestations of sincerity, integrity and courtesy on the one hand, and, on the other, a total incapability of observing the most sacred promises. It is an old story, the same which confronted Wellington in the early nineteenth century. The Portuguese is very like the African ; you despair of curing him of his weaknesses—which are, after all, seldom intentionally vicious—and yet you love him, because his kindly nature compels you.

The chief interest for the British public in Portuguese colonies arises from two distinct causes—financial interests and treaty obligations. Our financial interests are not large ; they involve certain railway schemes, the supply

of labour for the Transvaal mines, and a few plantation
and merchant enterprises. Our treaty obligations,
binding us in definite alliance with Portugal, may at any
moment involve Great Britain in a grave international
situation. The value, or otherwise, of such an alliance
is open to a difference of opinion. It is, however, imperative that our ally should observe all moral standards
which the civilized Powers are pledged to maintain with
all the forces at their disposal. Travellers, consuls,
merchants, sea captains and government officials have
repeatedly called attention to the prevailing slavery and
slave-trade in Portuguese West Africa ; both of which
detestable practices are in gross violation of Anglo-
Portuguese treaties, the Brussels and Berlin Acts. All,
or any, of the civilized Powers can at any moment—and
in point of responsibility should—intervene and demand
the abolition of slavery in Angola, San Thomé and Principe,
and if Portugal continued to beg the question by calling
slavery by some other name, that Power, or those Powers,
could, if they so desired, shake her out of her indifference
by casting the anchor of a battleship in sight of the ports
of San Thomé and St. Paul de Loanda. I am not advo-
cating such a course for one moment, but it is vital that
the British public should realize that in the event of any
Power signatory to anti-slavery Conventions waking up
for any reason, disinterested or otherwise, to treaty
obligations, and making some effort to discharge those
liabilities, such Power would be at once confronted by the
possibility of Britain's navy defending Portuguese
colonies, although run by slave labour. A pretty spectacle
indeed, Britain's matchless fleet defending the slaver,

COCOA CARRYING, BELGIAN CONGO.

ENTRANCE TO COCOA ROÇA, PRINCIPE ISLAND. (PORTUGUESE.)

only wanting old Jack Hawkins on the Bridge, to complete the picture !

Portugal shares with every other West African Power the problem of shortage of labour and with it the short-sighted energy of the impatient employer, who, beyond the ken of the official eye, frequently resorts to illegal means for increasing his supply. Domestic slavery survives in Portuguese Angola as well as in Nigeria, and in Belgian and French Congo. One can only estimate very roughly the slave population of Africa, but probably not less than a million human beings are to-day ignorant of the blessings of personal liberty. Mr. Nevinson, in his admirable book on " Modern Slavery," says of Angola alone, " including the very large number of natives who, by purchase or birth, are the family slaves of the village chiefs and other fairly prosperous natives, we might probably reckon at least half the population as living under some form of slavery." We cannot acquit many Powers in Africa from the charge of profiting adminis-tratively from this form of human chattelage, but when Portugal sets up a *tu quoque* plea we are compelled to differ. The dividing line between the Powers is that whilst many of them profit by this practice occasionally and for restricted periods, the Portuguese descend to the lowest level, adopt the native practice themselves and thus become not the " hirers," but the owners. In this way they endeavour to meet their interminable shortage in the labour supply. To what lengths they are prepared to carry this system may be gathered from the report of Professor A. Prister in the *Hamburger Fremden-Blatt* for 28th July, 1906 :—" In Angola, even in San Paolo de

Loanda, under the eyes of the Governor, the Bishop and the high officials," he alleges, are to be found "regular ' bridewells ' for the production of slaves." One of these, he says, he visited on the estate of " one of the richest Portuguese," sixteen miles from Loanda. There he saw a large number of women, with only a few men, at work. " Each woman has a little hut, in a courtyard enclosed by a wall, in which she lives with her young ones. The woman is always pregnant, and carries her last child on her back, during work, in Kaffir manner. The overseer of this plantation, who treated me in every respect with Portuguese friendliness, and took me for a great admirer of his breeding establishment, told me that about four hundred negroes were there, and added with a laugh that he had over a hundred young ones in the compound. This is just as if a cattle-breeder were boasting of the fine increase in his herds. When the young one is so far grown up that he can be put to some use, at from six to eight years of age, he enters into a so-called contract, or he steps quite simply into the place of a dead serviçal. For instance, Joseph is told that his name is no more Joseph but Charles, and immediately the dead Charles is replaced. He never fell ill ; he never died ; he only lives a second life." It is to be hoped that such incidents are rare even in Angola, but it brings home forcibly to the British mind the sort of colonies the " matchless navy " of Great Britain may be called upon one day to defend.

Certain apologists of the Portuguese are very fond of comparing the British indentured labour system with labour conditions on the Angolan mainland and the islands.   The labour system of the East and West Indies are by no means ideal, but there is a world of difference, not only in the daily management of this labour, but fundamentally.

In San Thomé the contracted labourer from Angola is a slave : he calls himself a slave, and the Mozambique free man holds him in contempt as a slave ; either he was captured, or purchased on the mainland with cash by the plantation owners just as men purchase cattle or capture wild animals.   Every single slave with whom I spoke, both on the mainland and on the islands, gave me the clearest account, replete with convincing detail, of the manner in which he or she had been either kidnapped or purchased.   Not a few of the slaves had " changed hands " several times before the ultimate sale to the planter.

The Slave's Case

In the back streets of Angolan ports, on the highways of Lobito and Benguella, and in the shady by-paths of Catumbella, the traveller may at any time penetrate the secrets of the tragedies which attach themselves to the souls of men and women who have lost their freedom. The same tragedies but with attendant secrets darker still, are locked within the breasts of the slaves on the Portuguese cocoa islands in the Gulf of Guinea.   There by the roadside, on the banks of crystal streams, up in the cocoa roças, and along the valleys thick with cocoa-trees, the traveller has abundant opportunities for penetrating the secrets of the miserable slaves.

Behind the mountainous coast of Angola, the town of

Novo Redondo hides itself in a hollow, as if ashamed of its history, or perhaps so that its traffic in human beings during past centuries might escape the attention of watchful cruisers. There, amongst a group of slaves and freemen, I met a woman with a story more eloquent than others because it was also so recent, so vivid and so forceful. She had not been long on the coast, for only a few months ago she had for the first time witnessed the Atlantic breakers tossing themselves with their impetuous fury on that strip of rocky shore. The hour was that of the mid-day rest, and the woman was sitting sadly apart from the other labourers. A glance at her attitude, coiffure and other characteristics rendered her a somewhat singular figure in that group of serviçaes, still there was a familiarity which surely could not be mistaken— somewhere in Central Africa those cicatrized arms, that braided head, had a tribal home.

" True, white man, I have come from far ; from the land of great rivers and dark forests."

" How were you enslaved ? " I asked.

" They charged me with theft and then sold me to another tribe, and they in turn to a black trader. This man drove me for many ' moons ' along the great road until a white man at D—— bought me and sent me here."

" Where am I going now ? Who can tell ? I suppose I shall be sold to a planter."

There was no need of the slave's reiterated assertion that she had been nearly ten months marching down to the coast ; the locality of her tribe was plainly set forth on the forearm by the indelible cicatrizing knife of her race. The journey from the Batetela ' ibe of +he Congo to the shores of Novo Redondo cannot be much less than 1,500 miles.   This was one of the most recent cases we discovered and shows that the slave trade in Portuguese territory is a question of the moment.

Fifty years ago, it is said, a ragged urchin ran the streets of San Thomé, holding sometimes for a five *reis* piece, sometimes for as many kicks, the heads of mules and horses for the affluent slave-planters of that island.   That ragged urchin to-day possesses a mansion in three capitals of Europe, and a stately car rushes to and fro with the sovereign lord of some thousands of slaves.   The sycophants, time-servers, and others of the crowd of parasitic admirers, who cluster round this august person, care little for the misery beneath that sordid splendour.   His wretched slaves spend their days from 5.30 in the morning until sunset cultivating cocoa, that their master may fare sumptuously every day in Europe, and finance dethroned Royalty which is not ashamed to use these ill-gotten funds in half-hearted endeavours to regain a discredited crown.   The slaves know nothing of this ; one thing only they know is that when the bell rings at sunrise they must devote their energies to the production of the cocoa bean until sunset, and that this weary monotony has in it not a glimmer of hope of cessation.

Along that picturesque road, known as the Mother of God road (philosophers might give us some reason why the slavers in all history annex the Holy Virgin), we once met a group of slaves with a sadness written on their faces which seemed almost to cry out, " We are lost souls."

" Are you well fed ? "

" Yes, white man, we are fed."

" Housed ? "

" Yes."

" Are you freemen ? "

" No, we are only slaves."

" Would you like your liberty ? "

" Aye, would we not, but Master won't liberate us."

Amongst that group was one old man quite grey, who declared he had been on the islands over thirty years, and his conversation so interested me that I asked him to describe his journey to the coast. This, though a story over thirty years old, was full of terrible interest. The old man had by this time gained some confidence, and when speaking of the district where he was first sold I became convinced that his home was in the far hinterland of the Congo. With unexpected suddenness I startled him by uttering one of the rhythmic morning greetings of his native tongue. The old man started at first, as if struck with a whip, then, like a man half awake, he appeared to reach after some unseen thing ; then at last it suddenly broke in upon him that the language he had heard was the music of his boyhood ; his wrinkled old face was wreathed in smiles, his tired eyes lit up, and then in short animated sentences he poured forth question after question.

" Oh ! white man, tell me about Luebo, tell me about Basongo."

" Tell me is Kalamba still alive ? "

The impetuosity of the questions, the lively gestures, the hungering look in those brown eyes showed how the old man thirsted for information of his little village away on the banks of the broad Kasai.

SLAVES ON SAN THOMÉ.

DISUSED SLAVE COMPOUND IN REAR OF HOUSE, CATUMBELLA.

The island of Principe has a horror all its own, for it is infested with the dread sleeping sickness. Conditions are so bad that the Portuguese dare not send the free labourers from Mozambique, lest their current of labour from that part of West Africa should take alarm and cease. White men and women are fleeing from danger, but the authorities still keep slaves within biting distance of the fever impregnated fly. Dr. Correa Mendes, a courageous Portuguese medical authority, has urged that every living animal should be killed as the only hope of saving Principe ; but none have yet dared to propose the liberation of the slaves.

The slaves of Principe present an even more melancholy appearance than do those of San Thomé. They appear to possess an instinctive knowledge that they are confined in a death-trap, and their appeals for liberation are piteously violent.

I cannot readily forget a conversation with four young slaves on Principe. Of these, but two had known freedom ; the others had been born of slave parents. On the features of one, the traces of sleeping sickness in an advanced stage were plainly marked, and though still labouring at his task, it was plain that death had already marked him for its own.

I asked the usual questions.

" Are you well fed ? "

" Yes, Senhor."

" Clothed and housed ? "

" Yes, Senhor."

" You are not flogged or beaten ? "

" Oh ! are we not ! "

" But I am told the planters never beat you."

" Tell me then, Senhor, how was this deep wound caused ? "

In support of this statement the whole group of slaves chimed in with exclamations and assertions that they were constantly flogged and beaten.

" Do you desire your freedom ? "

" Senhor, why taunt us ? Did you ever know an African who did not love his home and country ? "

" Well, I think there are people in Europe who will endeavour to emancipate you."

" Senhor, I fear when you get on yonder ocean, you will forget the poor slaves of Principe and San Thomé ! "

This latter reply was uttered with so desponding a note that I ventured to make the slaves a promise, which British honour—no less than British responsibility— should see fulfilled.

" Listen, I am now going to Europe and shall soon meet the liberty-loving British people. I know how they detest slavery ; I know how they will struggle for your liberty. Take this promise yourselves—and pass the word round the plantations to the other slaves—God helping us, we will set you free within two years."

The effect of this promise was good to behold, the eyes brightened, there was an elasticity in movement and grateful word of thanks as the slaves resumed their never-ending task. Even the slave in the fell grip of sleeping sickness appeared to share in the joy of a freedom he could not hope to experience.

Not all the slaves are purchased for plantation work, as the following typical instances will shew. Beautiful black women have their price. The day was indeed a hot one as I strolled along the shores of the Atlantic below the mouth of the Congo, when a finely-built young woman met me. Originally captured over 2,000 miles away, her fine figure and bright features had obtained for her captor a high price as " domestic " for the white man. To him it was nothing that she longed to exchange her captive life for that home away in the far interior, or that the roar of the waves was a perpetual reminder of the gentle lappings of the lake shore of Tanganyika. The woman was his slave, purchased with " honest money " —his slave until he ceased to want her, and then—well, he would sell her to the nearest planter and buy another, for healthy young girls are always marketable not only in Portuguese territory but in other parts of West Africa.

Another day two white-clad European travellers might have been seen moving in and out amongst the villages outside a Portuguese town of Angola, exchanging greetings with half-dressed natives. Presently it is realized that this is no casual visit of curious strangers, for it is obvious that the white man's handshake is but an excuse for a closer scrutiny of the arm, the temple, or the chest, and the natives gather round the travellers as they proceed from group to group. Something now arrests attention, for the white man is sitting down amidst a party of four or five women.

In a few minutes confidence has been gained, and the women submit to an examination of certain marks cut years ago on their arms and foreheads. The white man first tries a sentence in a tongue unknown to the group of interested onlookers, but there is no response from those to whom it was addressed. He tries another, and there is a sudden silence ; all eyes are directed to a woman who, after a faint cry of amazement, is gazing fixedly into space, for the white man had by that sentence struck a chord silenced by long years of sorrow and suffering. The woman gazed on silently and intently as if trying to recall a half-forgotten past. She travels in thought back over yonder mountains, across the hot plain, and on by rippling streams and through valleys thick with ripe corn, away across the Cuanza river, on for months to Lake Dilolo, where she sees again as in a vision the white man who bought her from the native slave-trader. In fancy she leaves the cornfields of Angola, crosses the upper Kasai, and is away north beyond Lusambo, and westward to the little Congo village with its deep green plantain groves and manioca fields.

A remark breaks the spell, and she realizes that it was but a dream, for she is still a captive ; but the white man speaking her native tongue is no dream—he is still there speaking the language that sounds like the far-off music of another life. The light of hope dawns in her eyes as she turns on the traveller, pleading, " White man, can't you take me home ? "

It must not be supposed that all the San Thomé planters on the island believe in, or defend, present conditions, any more than it must be supposed that, without exception, they are habitually guilty of inhuman maltreatment of the slaves. The charge of maintaining slavery most of them emphatically deny, and in support of their contention point to legal contracts which cover the original

SLAVES ON COCOA ROÇA, PRINCIPE ISLAND.

THE END OF THE SLAVE. TWO SLAVES CARRYING DEAD COMRADE
IN SACK TO BURIAL.

transaction by which the labour was obtained. They also remind the investigator that the labourers are paid. There are, however, some honest planters who admit that the original " contract " was not altogether genuine, and the statements made by the planters and the slaves respectively with regard to the wages paid, differ so absurdly that one is compelled to dismiss both.

To many managers definite acts of cruelty would be highly repulsive. It is furthermore very obvious that not a few owners and planters do everything which science and money can provide to make the lot of the slave a happy one. The planters argue with much warmth and sincerity of conviction that the labourers are better housed, fed, and clothed on the plantations than they would be in their mainland villages. Their melancholy demeanour and their insistent desire for liberty, the low birth rate and frightful mortality amongst the slaves is put down very largely to the gross obstinacy and stupidity of the enslaved negroes.

If the planters are questioned upon the desire of the slaves to regain their liberty they reply that this would be an act of injustice because many of the labourers have forgotten the districts from which they were originally " recruited " and that even if complete repatriation were carried through the men and women repatriated would probably fall a prey to evil influences on the mainland.

The attitude assumed by the Portuguese authorities towards the question of slavery in their West African colonies has hitherto been first of all one of inferential denial that slavery exists, and secondly they call attention to the elaborate regulations framed for protecting the natives from any infringement of their liberties.

On paper, the labourers are contracted for short periods of service in Angola and the cocoa islands ; are said to have a happier lot than any other contract labourers in the world ; and that any who so desire are free to return to their homes at the termination of their contracts. A great deal more is on paper which, if practices only accorded with the minimum of professions, would assure the cessation of slavery in Portuguese West Africa.

Perhaps nothing written in the earlier days upon this question has brought out so forcibly the " ownership " feature of labour conditions as the disclosures made in the Cadbury—*Standard* libel action. In that trial Sir Edward Carson called attention to a circular forwarded to Messrs. Cadbury referring to the sale of an estate in San Thomé. The stock enumerated included one item, " Two hundred black labourers . . . £3555." This gives the average price of the slaves as £18 *per capita*, taking the sick with the healthy and the (young with the old. Various prices are quoted as the value of the slaves, but this depends, of course, upon physique, sex and age. Mr. Joseph Burtt, the Commissioner of the cocoa firms, gives £25 to £40, whilst Mr. Consul Nightingale stated £50 as the average price. When in Portuguese West Africa several of the slaves were even able to tell us the prices at which they were purchased by the different middlemen, and occasionally even by Portuguese themselves.

The evidence now to hand of the existence of both the slave-trade and slavery is overwhelming.

On November 22nd, 1909, the Portuguese Foreign

Minister called upon Sir Edward Grey, apparently with the object of discussing this question, and in conversation the Foreign Minister informed M. Du Bocage that the information he " had received from private sources placed *beyond doubt* * the fact that it had been the custom for natives to be captured in the interior by people who were really slave-dealers ; the captured natives were then brought down to the coast and went to work in the Portuguese islands."

On the 26th of October last, Sir Arthur Hardinge, whose intimate knowledge of slavery questions is probably unequalled, informed the Portuguese Minister for Foreign Affairs that when he was in Brussels, he " had heard serious complaints in official circles at Brussels of the way in which slaves were kidnapped by Angola caravans from the Kasai district of the Congo, which shewed that the charges made did not emanate solely from missionaries or philanthropic sentimentalists."

In July, 1909, an exhaustive series of regulations were issued from Lisbon. The 139 articles covered almost every actual and conceivable feature of the whole labour question, but as Mr. Consul Mackie pointedly remarked—

" The Angolan native . . . is contracted in a
" wild state under circumstances of doubtful legality,
" and is so convinced that he is a slave that nothing
" short of repatriation, which should therefore be
" compulsory, would serve to persuade him that, at
" least in the eyes of the law, he is a free agent. It
" would obviously be useless to argue that the

* Italics mine.—J. H. H.

" ' serviçal ' is not a slave merely because he is
" provided with a legal contract, renewable at the
" option of his employer, in which he is officially
" proclaimed to be free."

The evidence, therefore, that the Portuguese colonial
labour systems are pure slavery is confirmed (a) In a
British Law Court, (b) by the British Foreign Minister,
(c) by the British Consul on the spot, and (d) by Sir Arthur
Hardinge.   Could anyone desire more emphatic evidence
than is now provided ? and this does not exhaust the
available sources, for even the Portuguese themselves
have now been forced to admit that the slave-trade is
very much in evidence.

Writing to Sir Arthur Hardinge on October 23rd last,
the Portuguese Foreign Minister admitted that there
was—

" Slave-traffic with the inhabitants with Luando,
" in the district of Lunda. . . . It was ascertained
" that in reality the Bihean natives were in the habit
" of settling their debts and disputes by means of
" ' serviçaes.'   Two convoys proceeding from Luando
" were captured, and the serviçaes handed over to
" the delegate of the Curator concerned, to be re-
" tained until claimed by their relatives, to whom
" the necessary notice was sent."

It is instructive to note that the Portuguese Minister
in this passage makes no distinction between serviçaes
and slaves, yet when unofficial critics declare that in
practice the terms are indistinguishable, they are con-
demned for deliberately confusing the public mind.   It
is also of importance to bear in mind that the Lunda

Province is included in the territories which come under the operations of the Berlin Act, whereby every European Power is under the most solemn responsibility to secure throughout these territories the abolition of slavery.

There is further the evidence, from a Portuguese source, in the fact that in 1911 eleven Portuguese are reported to have been expelled for engaging in the slave-traffic. The Governor of Angola informed Mr. Drummond Hay that nothing severer than the order of expulsion was administered owing to the lack of " conclusive evidence," but four months later the Portuguese Foreign Minister admitted that the Europeans " by the inquiry were found guilty of acts of slave-traffic." Sir Arthur Hardinge pointed out that

" the 5th article of the Brussels General Act con-
" templated severer penalties in the case of persons
" engaging in the slave-trade than an order of
" expulsion before trial or a prohibition to return
" to the colony, which such persons, if convicted of
" a serious criminal offence there, would hardly
" need."

When, therefore, the long stream of unofficial testimony upon the existence of the slave-trade and slavery in the Portuguese colonies is confirmed in turn by the British Minister at Lisbon and by the British Consul of Angola, and moreover when Sir Edward Grey, who always chooses language with exceptional care, officially informs civilization that the charges are proved " beyond doubt," and finally when the Portuguese authorities are driven to admit it, then surely the time has come to cease gathering evidence and to set about some substantial and far-reaching measures of reform.

To meet a situation which is a grave international scandal and a potential menace to European peace, what do the Portuguese offer to civilization ? They claim good treatment of the labourers, humane regulations and repatriation of the slaves.

It is common ground that the slaves upon the islands are, generally speaking, well fed, housed and fairly well clothed, but the slaves themselves are emphatic in their assertions that they are frequently beaten. This the Portuguese deny, but those who know West Africa are perfectly well aware that it is impossible to keep something like 40,000 slaves working on the cocoa farms at the rate of " a man per hectare " without a considerable amount of " pressure."

The regulations are exhaustive upon every feature of the labourer's life except emancipation, from the time he is " recruited " until he is buried, but as Mr. Consul Mackie has recently pointed out, " The absence of any (regulations) for the return journey in the event of the labourer declining to accept the conditions of the contract is somewhat suspicious." That the keeping of statistical records is part of the ordinary administrative routine is a common-place, but that this elementary duty may be performed, a regulation was issued three years ago, yet to this day that instruction has never been carried out, with the result that no reliable information is possible with regard to the birth and death rates.

According to regulations, every labourer's history is carefully recorded and the fullest details endorsed upon the contract, yet in the case of 135 out of 163 slaves repatriated in the early part of last year it was impossible to state how long they had been upon the islands. This is not only a further evidence of the futility of Portuguese regulations, but it constitutes additional evidence as to the fictitious nature of the supposed " contract systems."

To the tragedy of slavery on Principe is added the ever present horror of sleeping sickness which is everywhere raging on the island. To the lasting disgrace of the Portuguese Government it still permits the retention of slaves on the island where conditions are so bad that, as I have already pointed out, the eminent Dr. Mendes has advised the killing of all the cattle on the island in the hope of checking the ravages of the disease. Here again the Portuguese Government has been content to meet the situation with paper " regulations," which would be comic if it were not for the distressing condition of these wretched slaves. According to these regulations all the slaves " must wear trousers to the heel, blouses with sleeves to the wrist, and high collars . . . they must wear on their backs a black cloth covered with glue ! " It is barely necessary to state that these regulations are openly disregarded in every particular.

Finally the Portuguese point to their highly-regulated system of repatriation. Sir Edward Grey and Sir Arthur Hardinge have emphasized that " one excellent test " of a desire for reform " will be the rate and method of repatriation." Since the year 1888 some 67,614 slaves are known to have been shipped to San Thomé and Principe from the Angola mainland, but it is also known that a good deal of smuggling has been carried on. Then, too, there are some slaves born on the islands prior to the year 1888, and these are claimed as the property of the planters. At the same time the death rate has been very high and the birth rate extremely low, with the result that the estimate of the slave—or according to the Portuguese the Angola serviçal—population of about 37,000 is probably fairly accurate.

In 1903 a repatriation fund was established with the object of providing the repatriated slaves with a sum of money upon their landing again on the mainland, and there was a further complicated arrangement which could have no chance of being effectively carried out, whereby "recontracted" slaves should receive their bonus in 6 per cent. instalments every quarter. This was no philanthropic contribution, but actually represented a regular deduction of 50 per cent. from the "wages" of the slaves and serviçaes, until May, 1911, when the deduction was raised to two-thirds, leaving the labourer only one-third, and as most of the slaves appear to die prematurely, the benefit they receive from their "wages" is a negligible quantity.

From the year 1903, when this fund was instituted, until 1907 these deductions from wages were actually left in the hands of the planters. In December, 1907, they admitted to holding £100,000—a not inconsiderable capital fund for working their plantations. In 1908 the fund was transferred to the Government bank, but in the autumn of that year it had by some means, yet to be discovered, shrunk to about £62,000. After about nine years' working this fund stands to-day at approximately £100,000.

In Sir Edward Grey's despatch to Sir Francis Hyde Villiers of November 3rd, 1910, he pointed out that public opinion would be favourably impressed with a "regular and satisfactory" method of repatriation. During the early part of this year 900 slaves were repatriated to Angola and 500 more were due to leave at the end of June. Probably, therefore, it may be safely estimated that since 1908 something like 2000 will have been returned to the mainland before the close of this year, but even so this still leaves something like 35,000 still in slavery. Though this "rate of repatriation" shews some improvement, it cannot be regarded as satisfactory in view of the fact that unless it is materially accelerated, it will take not less than twenty years to liberate the whole of the slaves on the two islands.

Turning to the method of repatriation, it is clear that this is being carried on in the most inhuman and barbarous manner. When we were at Benguella, the condition of the repatriated slaves was so distressing that we offered the Governor a sum of money to provide the miserable creatures with food and medicine. This His Excellency could not receive, nor could he allow the Curador to accept it.

The planters, realizing that civilization demands that "repatriation" should take place, are just now permitting it, but they are at the same time doing everything in their power to discredit it. There is some evidence that they are only liberating the sick and worn out, for of twenty-eight slaves recently liberated, whose ages were known, Mr. Consul Drummond Hay tells us their average age was forty-two years, and their period of labour on the islands averaged thirty-one years.

The Portuguese journal *Reforma* in its issue of August 19th, 1911, exposes in convincing language the condition and real objects of this so-called repatriation :—

" The greater part of these ' serviçaes ' were put " on board without being told what their destination " was, and without any money.

" The first batches that came here had some " money—one of them had 150,030 *reis* (£30)—but " the last lots arrived without any ; thus some were " expatriated instead of being repatriated.

" These people did not bring a single penny, and " it was through charity alone that they received " food and shelter. Almost every one of these " unfortunate people, who have done twenty years " of hard labour, arrived in ruined health, and some " of them died shortly after their arrival.

" It is probable that this form of repatriation is " a stratagem which will, on account of the protests " that will be raised against it in this province, " enable the planters to argue that repatriation is " unproductive of any good results, and that the " truth is, as they have said all along, that people " who have once gone to those islands never want to " leave them again, well knowing that they could " not find a better spot in this world."

The Brussels Conference of 1890 foresaw this danger and made provision for it in Articles 52 and 63. The latter stipulated that :—

" Slaves liberated under the provisions of the " preceding Article shall, if circumstances permit, " be sent back to the country from whence they came.

"In all cases they shall receive letters of freedom "from the competent authorities and shall be "entitled to their protection and assistance for the "purpose of obtaining means of subsistence."

Portugal was signatory with other European Powers to the Brussels Act.

Further evidence of the inhuman manner in which the slaves are repatriated appeared in the Portuguese journal, *A Capital*, of June last. The writer, Hermano Neves, reported that an officer on a Portuguese ship informed him that in one trip there were 269 liberated serviçaes, that of these unfortunate beings landed at Benguella only one was given any money ; the remainder, unable to obtain employment and without money to buy food, were left to starve. " A few days later, there lay in the outskirts of Benguella, out in the open, no less than fifty corpses ; those who did not or cared not to resort to theft in order to live had simply died of starvation."

How comes it that in spite of endless " regulations," almost every line of which boasts humane sentiments, and of a Government in Portugal which blazes upon the housetops its devotion to the cause of human freedom, these deplorable conditions prevail in the West African colonies ? The reason has been advanced without any equivocation for years, namely :—the Portuguese colonies are out of control. Portugal may send a shipload of regulations out of the Tagus every week and the planters will welcome them—as waste paper. Some of us have said this for years and have suffered the inevitable abuse, but with the publication of the recent White Book, we actually find this contention corroborated by the Portuguese Government. Senhor Vasconcellos, replying to Sir Arthur Hardinge's representations upon the abuses, admitted that :—

"The Governors whom he had sent out to give "effect to its (the Government's) instructions had "been to a great extent paralysed by the power of "vested interests."

This is, of course, obvious to all those who realize the inner meaning of the fact that within ten years the cocoa islands have had something like twenty-five Governors. An admission of this nature by the responsible Portuguese Minister goes quite as far, if not farther, than the most extreme critic of Portuguese colonial administration.

The existence of slavery and the slave-trade now corroborated by officials of the first rank in London and Lisbon, supported by Consuls, and now by the Portuguese themselves, leaves no longer any need for unofficial persons to spend further efforts in an endeavour to establish the fact. Sir Edward Grey's "beyond doubt" is in itself sufficient for the great mass of sane men. With the breakdown of the Portuguese regulations and the violation of international treaties, coupled with the Portuguese admission that their colonies are out of hand, what can be done to set free the slave in San Thomé and Angola ?

The question is frequently asked what would be done if the slaves were set free. We are told that to dump down on the West Coast of Africa 40,000 penniless slaves originally drawn from homes in the far hinterland, might involve great hardship.   We all agree, for we now know that thousands of slaves were obtained for Portuguese colonies from Belgian Congo, through the help of the revolted Congo State soldiery—a body of men numbering according to circumstances, from 1000 to 5000, who only kept up their rebellion by purchasing slaves with arms and ammunition from the Portuguese half-castes and natives.   Again, no one can read the thrilling story by Colonel Colin Harding in "Remotest Barotseland" without being convinced that the Portuguese obtained many slaves from British territory.   Lake Dilolo, the greatest of all the slave-markets, is but a comparatively short march from Rhodesia, and, in view of local conditions, it is inconceivable that natives of British Rhodesia have not been drawn into the slave-traders' toils in that region. This feature has recently received an emphatic confirmation from Mr. F. Schindler, a missionary of over twenty years' experience in Angola.   He writes :—

"I have seen thousands of slaves coming from "the Belgian Congo and Rhodesia being taken "westwards by Bihean slave-traders and in some "cases by half-caste Portuguese, and both by their "tribal mark and by their speech I had no difficulty "in recognizing them as belonging to tribes that are "not found in Angola."

How many of these slaves were, and are, enslaved on the mainland, and how many ultimately found their destination to be the cocoa islands, it is impossible to say, but we do know that generally it was the hinterland native which the slave-traders shipped to San Thomé and Principe.   Moreover, some of us have seen these people

of the Batetela and Kasai tribes on the roads and plantations of the islands, the cicatrized arms, legs, chests and
backs plainly indicating their origin.

The first essential, therefore, is that of determining the
countries of origin of the slaves on the islands.   To whom
can this task be assigned ?   Obviously not to the planters ;
it might be entrusted to a disinterested Portuguese
Commission, but others have responsibilities and vital
interests—Great Britain and Belgium would both possess,
if not the right of membership, certainly the right to
watch proceedings on behalf of any natives whom they
had reason to believe had been obtained originally from
British or Belgian colonies.

The planter holds that the slaves are happier on the
islands than they could ever be on the mainland ; this
interested and *ex parte* statement cannot obviously be
accepted as final. The native, and the native alone,
should be allowed to determine his, or her, destiny. I
admit it is conceivable that a few slaves, for various
reasons, would elect to stay with their owners, and no
compulsion should be put upon such to leave the islands,
but beyond all question the majority of the 37,000 slaves
have a deep-rooted and a passionate desire to return to
the homes of their birth.   When visiting the cocoa islands
in October, 1910, Mr. Consul Drummond Hay sent his
interpreter amongst the slaves to ascertain whether they
desired their liberty, and in his report to Sir Edward Grey
says : " My interpreter went among the Angola ' serviçaes '
and his inquiries as to whether they wished to be
repatriated were mostly answered in the affirmative."
This, be it remembered, was said by the slaves on what

are admittedly the show plantations. Take these slaves aside and engage them in conversation, and before many minutes have passed, the appeal will involuntarily burst forth, " White man, give us our liberty ! "

Having ascertained the districts of Central Africa of those who desire emancipation and a return to their villages, it should then be the duty of the representatives of Portugal, Britain and Belgium, to see to it that their respective subjects are quickly and safely returned. Much has been made of the difficulties which would attend any schemes of repatriation, but in many quarters these difficulties have been purposely exaggerated. Given an honest desire to repatriate, the task would at once become simple. Take first the Angola natives. The Portuguese could, if they chose, send them back in batches of 50 or 100 for a given district ; a body of such dimensions attaching itself to an up-country caravan, travelling under official protection and possibly with a small escort, would present too solid a company to permit of attack. Moreover, officials, traders and missionaries, might all be notified of such companies journeying from the coast and instructed to aid them as far as possible. The Lobito—Katanga Railway Company would doubtless be willing to give cheap passes to batches of slaves originally secured from the different centres through which its line now passes. It would be distinctly to their interest to do so, apart from humanitarian considerations.

We now know that providing the Portuguese Government would set at liberty the slaves originally captured from the upper reaches of the Kasai, the Belgian Government is prepared to send ships to San Thomé to carry them back to the Congo, transfer them to steamboats which would take them back to their homes, or at least within a day or two's march. This journey could now be accomplished in less than a month, whereas several of the slaves obtained from Belgian territory informed us that their original journey in the chain gang to the coast had involved a tramp of considerably over òne year. There is reason to believe that not only would Belgium undertake this task, but she would do so without requiring any financial return whatever.

The third and probably the smallest section of the slaves on the islands—British subjects—can assuredly present no difficulties. Great Britain could with the greatest of ease collect her slaves at San Thomé and transfer them to Rhodesia and Barotseland, via the Cape.

Portugal should be invited to send an international commission to West Africa, composed principally of Portuguese, but with a British and Belgian element, assisted by men experienced in the tribal languages and cicatrices of the hinterland peoples. This commission to be empowered to investigate the whole question and to issue freedom papers to all slaves appealing for liberty. In view of the advertised hatred in which the present Portuguese Government professes to hold every form of servitude, such commission might easily be appointed in friendly co-operation with the Powers primarily concerned. If this were done, the Portuguese Government and nation would at once merit and undoubtedly receive the warm appreciation and support of the civilized world.

If, however, the Portuguese Government, after

admitting their incapacity to control their West African colonies, refuse the co-operation of friendly Powers and maintain a system of labour which violates in several respects international treaty obligations, it is obvious that, however much Great Britain may regret it, she cannot continue an Alliance which may at any moment involve her in a position of the utmost gravity.

It would be idle to overlook the extremely serious nature of the statement made by Sir Edward Grey in the House of Commons on April 3rd, 1912. The Foreign Secretary then declared that the defensive treaty of alliance between Great Britain and Portugal, though it had not been confirmed since 1904, was, like all similar treaties which, " not being concluded for any specified term, are in their nature perpetual."

Thus it would seem that if any one or more Powers signatory to the anti-slavery clauses of either the Berlin or Brussels Acts, should awake to their clear rights and solemn responsibilities and proceed by any show of force to insist upon the abolition of slavery and the slave-trade in Portuguese colonies, the maritime and land forces of Great Britain could under this Alliance be forthwith summoned to protect these Portuguese colonies against the " Aggressors."

There are some things impossible to the strongest of Ministers, and the Portuguese Government must realize that the British people, however much they might desire to do so, cannot allow the continuance of an Alliance with a Power which by persistent violation of international obligations exposes not only herself, but her ally, to a defence of slavery and the slave-trade. Now is the time for Portugal to accept the friendly advice and help of Great Britain, but as Mr. St. Loe Strachey has recently said :—

> " *Either the Portuguese must put an end to slave-*
> " *owning, slave-trading and slave-raiding in the*
> " *colonial possessions which we now guarantee to them,*
> " *or else our guarantee must at once and for ever*
> " *cease.*"

IV

THE FUTURE OF BELGIAN CONGO

BELGIUM for the time being is in the saddle, but for how long ? Will she prove strong enough, wise enough, great enough to bring order out of the chaotic state of affairs into which her late ruler plunged the Congo territories ? It would require a bold man to give an unqualified affirmative to this question. Cover several thousand miles of that territory, live for months with the aboriginal tribes, discuss administrative problems with Congo officials, watch the operations, and listen to the conversations of the German and Portuguese merchants —and a permanent Belgian control of the Congo becomes a matter of considerable doubt.

Belgian Congo, the largest single political division of Africa—French Sahara alone excepted—possesses land and climate of distinct features, and, properly administered, could pour into the European markets raw materials now demanded by many of our industries. The total area of the old Congo State was just over 900,000 square miles, or eight times the size of Great Britain and Ireland. A considerable proportion of the territory is covered by a series of gigantic swamps, with ribs of dry land and ironstone ridges dividing rivers and lakes. The whole of these low-lying territories are covered with thick forest undergrowth, which renders them impenetrable except along the native tracks. Throughout the Equatorial regions it would be extremely difficult to discover a single acre of open country, and in the territory covered by the Bangalla and its tributaries it is only with difficulty that even a camping ground can be obtained. Mobeka, the State Post at the confluence with the main Congo, was actually built by gangs of forced labourers carrying baskets of soil in an almost endless stream for a distance of nearly two miles inland. This post was formerly the head-quarters of the notorious Lothaire and it remains to-day a monument to the luxury with which he surrounded himself ; the carved woodwork from Europe, the doors and windows, and general upholstery are indicative of the high favour, or fear, in which this gentleman was held by King Leopold. Northward beyond the Aruwimi and southward of the Kasai the character of the country changes considerably. The eternal forests of the Equatorial regions give place to rolling veldt or open plains. Instead of swamps and marshes there are hills and valleys, although, unhappily, neither fertile nor occupied by a virile or extensive population.

For nearly a quarter of a century the Congo territories have suffered from uncontrolled exploitation. Twenty-five years ago the forests were thick with mature Landolphia rubber vines. This species of rubber is of very slow growth and probably some thousands of the larger vines extend over 100 years. Scientifically tapped in the season, this great vegetable asset would to-day have been almost unimpaired and the Congo could still have continued pouring forth 5000 tons of rubber per annum to

Europe. Nothing of the kind was attempted ; the stores of vegetable wealth carefully husbanded by nature for generations were exposed to ruthless plunder, the mad scramble for rubber at any cost to humanity and common-sense denuded the forests. The vine growths of a generation were hacked to pieces, and even to-day millions of dead fragments of vine may be seen scattered all over the hinterland forests. Even the roots were not spared, for the unhappy natives, driven to desperation by the white rubber collectors tore up the roots and forced them to disgorge their stores of latex. Rubber is still to be found, but in much smaller quantities, in the Aruwimi district in the north, the Lomame and Lukenya basins in the east, and also in certain districts in the Lake Leopold region, but no merchant should to-day enter the Congo with a view to making money from virgin rubber.

King Leopold knew all along what the Belgian Government now knows—that the greatest economic asset of the Congo would have disappeared by the time the Belgians inherited the colony, and he met the situation by the issue of two decrees : one instructing all agents and Government officials to lay down rubber plantations round every factory, and the other promulgating heavy fines and penalties for the severance of indigenous rubber vines. The latter decree was generally treated by whites and natives alike as an instruction " *pour rire* "—a fact known and probably anticipated by King Leopold. The instruction to lay down rubber plantations happened to meet to perfection a feature in the system of Congo State exploitation.

In those early days—from about 1897 to 1904—

there might be seen at every rubber collecting centre gangs of men, women and even children, chained or roped together by the neck, and these were the hostages which were being held by the " Administration " until a sufficiency of rubber had been brought in to redeem them. Generally these hostages were captured from amongst the old, the sick and afflicted, or even from the women and children, the object being to force the young and able-bodied into the forests to gather the rubber which would " redeem " the father, mother, sister or child.

The question which had hitherto confronted the officials was that of finding work for the hostages, for the Royal Rubber Merchant was known to favour every expedient which would strengthen the faith of the natives in the dignity of labour. The instructions, therefore, to lay down rubber plantations exactly met the situation, and the thousands of hostages throughout the Congo were forthwith set to the task of clearing forests and planting rubber. This removed from the wretched hostages their last hope of prolonged liberty, for it became doubly advantageous to capture and retain them. The slightest shortage of rubber was a sufficient pretext for capturing more hostages and thus provide labour for the plantations. A perfect equation was in this way maintained—if less rubber came in from the forests, more hostages would be laying down this new source of potential revenue. Tongue cannot tell, neither can pen portray the miseries involved in the laying down of these plantations, but the sight of the suffering natives can never be effaced from memory. The Congo chain gang respected neither position, age nor sex, sickness or health ; it held fast alike the old chief, the weakly man, the young girl and the expectant mother—a terrified mass of humanity trembling under the dreaded crack of the whips. The sentry overseers regarded them as the carrion of the Congo, for their relatives were guilty of the greatest of all offences, inability to satisfy the impossible demands for rubber. The infant in terror clung closer to the mother, as the woman winced under the lash of the whip. The young wife brought forth her first-born in her captivity and was left without any attention to battle with her weakness, or to succumb. To make a recovery was to resume her work of rubber planting within two or three days, with the new-born babe tied to her back. Darker deeds, too, were committed, and some rubber trees of to-day were literally planted in the blood of victims.

A writer, " Father Castelin," greatly impressed with the wisdom of this undertaking, but apparently caring nothing about its tragedy of human suffering, estimated from documents placed at his disposal that the " new source of revenue " which had been bequeathed to the Belgian nation, provided 13,000,000 rubber trees. This " new source of revenue " could hardly fail to provide an annual return of less than two francs a tree, thus assisting the budget with an asset of more than a million sterling per annum. This alluring prospect so impressed the new Belgian Colonial Minister that he added to his difficult and recently acquired administrative task that of rubber production on a " business basis."

When Monsieur Renkin introduced his famous Congo reform bill, it contained a proposal to extend the existing plantations by 50,000 acres. This in itself was a serious departure from recognized colonial principles in that it wedded the newly acquired colony, for better or for worse, to commercial undertakings. The whole enterprise from beginning to end is beyond question a miserable fiasco.

In our recent travels we have visited large numbers of these plantations. They are all of them characterized by neglect, the majority have been abandoned and are everywhere falling a prey to rapidly growing forest undergrowth. A considerable proportion of the trees, as if in protest against the violence which their planting involved, are now drying up from the roots. In spite of the millions of rubber trees planted in the Congo, many of these being more than ten years old, no plantation rubber has yet been profitably exported, nor is there any hope entertained by the officials on the spot that plantation rubber will ever be an economic success.

Inseparably interwoven with the exhaustion of the economic resources is the exhaustion of the people themselves and the break up of their social life. Stanley estimated the whole of the Congo population at something over 40,000,000. This was, of course, the merest guess, but probably the Powers at Berlin did commit to the care of King Leopold not less than half that number, i.e. 20,000,000. To-day the official estimate gives the total population at something under 8,000,000. It may be asked whether I should estimate that more than 12,000,000 of people perished under King Leopold's regime. I can only reply—certainly not less. The only ascertainable data upon which an estimate can be based would amply confirm such a statement. Many towns whose population

was known almost to a man twenty-five years ago have
disappeared entirely, and there is not one town to-day
but has lost over 75 per cent. of its population within the
last three decades. There is one redeeming feature, viz.,
that since Belgian occupation there is some evidence that
in several districts the appalling death rate and low birth
rate show signs of regaining a more normal standard.
This was the most apparent in the old sleeping sickness
areas, for we noticed that wherever the Belgian reforms
had been most completely applied, there the ravages of
sleeping sickness appeared to be more or less checked.

When Belgium annexed the Congo, she for many
months retained the old Congo State flag ; she still retains
the sobriquet " Bula Matadi " ; she retained, and still
retains many of the old Congo officials, and finally she
retained her interest in rubber. These indications did
not escape the notice of the natives who are never slow
to detect circumstantial evidence, to say nothing of the
enlightening influence of the old witch doctor ! The
consequence is that the natives distrust the new " Bula
Matadi " as much as they did the old one, for to many
of them there is no visible change. Thus Belgium finds
herself in possession of a colossal colony whose economic
resources are exhausted, whose population has been
seriously diminished, and whose native tribes everywhere
mistrust her administration.

The foregoing features present Belgium with a problem
not to be solved easily by the most experienced and powerful of colonizing Powers. International obligations, too,
cannot but make that task more difficult. The Congo
must still work out its salvation under the guardian

eye of the fourteen signatories to the Berlin Act.   It is
still the duty of each of these Powers to " watch over the
moral and material welfare of the native tribes."   Not
only so, but the Congo colony is further restricted by
separate treaties with all the Great Powers, which together
provide a shoal of difficulties through which it will not
be easy to steer the Administration without disaster.

The Congo territories, however, are not without assets,
which, in the hands of a bold statesman, are capable of
making Central Africa one of the greatest wealth-producing
areas of the Continent.

The first asset is in the riverine system of the Congo.
The main river has five large tributaries, each of which
provides from 500 to 1000 miles of navigable waterway ;
the Busira, for example, 200 miles from the mouth gives
no soundings at a depth of 1000 feet.   Each of these in
turn possesses numerous smaller, but still navigable
tributaries.   Altogether this fluvial system renders water
transport possible for over 10,000 miles, whilst for large
canoes and launches there is more than twice the waterway.   I know of no district, no matter how remote from
the great fluvial highway, which is removed more than
four days' march from a river bank.   In some parts of
the main river the width is considerably over five miles,
and in others it takes a canoe nearly half a day to thread
its way between the network of islands which cover the
river between north and south banks.   There is, however,
the outstanding drawback that as a commercial asset the
whole waterway is blocked at the mouth, strictly speaking
ninety miles from the ocean.   There the cataract region
begins which has hitherto defied engineering skill.

Between Matadi and Leopoldville, a distance of just over 350 miles, seven such natural impediments prove an insurmountable barrier to water transport. This distance is covered by a railway which connects the lower river with Stanley Pool, the upper river port. The line is undoubtedly a thing of beauty, but travelling on it is certainly not a " joy for ever," climbing up almost impossible slopes, skirting ravines and lightly circling mountain ranges—a triumph of engineering skill, whose construction, it is estimated, cost a life a sleeper. Its 2 ft. 6 in. gauge and its miniature rolling stock are, however, totally incapable of dealing with the potential transport of a colony more than half as large as Europe.

At present transport on the Congo railway is in hopeless confusion and the merchant is fortunate indeed whose goods occupy less than a month traversing that 350 miles, for the bulk of goods require six weeks to reach Leopoldville, the port of Stanley Pool, from Matadi on the lower river. When we were at Matadi there was still 1000 tons awaiting transport, a small task for European and American freight trains, but an entirely different matter on a line where we saw a Congo engine with twenty tons only in her trucks make no less than three attempts up an ordinary incline. The Congo railway, at present the only link between the ocean and the Upper Congo, presents to the Belgian Government a two-fold problem. The first question is whether it is possible to turn the whole track into a broad gauge, capable of bearing heavier rolling stock with reasonable safety—an initial problem of doubtful solution, and with it the second is coupled. If this line were practically rebuilt, at immense cost to the Belgian Exchequer, what reasonable guarantee has Belgium that for all time the French Government will refrain from constructing a railway from the seaboard of French Congo to Kwamouth on the confluence of the Kasai with the main Congo? Given such a condition, it is all over with the Belgian Congo railway. We know that many patriotic and far-sighted Frenchmen are seriously considering this proposition. Then, too, the French are great railway engineers, and I am informed that the physical conditions of the country through which such railway would pass are entirely good. If the French line were built, the Upper Congo would be brought at least five days nearer Europe for passengers and mails, while merchandise would probably save three weeks to a month in reaching its destination.

Even if Belgium provided an unchallengeable connecting link between the lower and upper reaches of the two fluvial systems, the Congo river is beset with political potentialities of no mean order. It remains to-day an international highway which presumably any five European Powers may, if they so choose, bring under the control of a five-Power river board of management. As an asset the Congo river is gravely depreciated by the topographical features from Stanley Pool to the mouth which place the whole Congo colony at the mercy of the Power which holds French Congo, and thereby the highway to the ocean.

Given security of control and also of communication, what economic future, actual and potential, is there for Belgian Congo?

That rubber of the indigenous kind exists to-day in

the recesses of the forest is true. This, as I have said, applies especially to the Aruwimi, Lake Leopold and Kasai regions, but only in comparatively small quantities. This indigenous product finds a sale to-day only because of the high prices which rubber has commanded during recent years. Many manufacturers are now refusing to touch native rubber at all, because it is so full of impurities. There are, indeed, many competent observers who state that when in a few years' time the yield of cultivated rubber, coupled probably with a successful manufacture of synthetic rubber has forced down the price of the better qualities, then the common and impure varieties from West Africa will be driven out of the market altogether. Of the various classes of rubber, that of the Congo is probably the worst, consequently the future of the colony cannot be based on an exploitation of the indigenous rubber latex.

Ivory has in the past figured largely in the Congo budgets, but the ruthless exploitation of rubber had its counterpart in the wanton destruction of elephants in order to obtain rapidly every tusk of ivory. The old Congo State agents frequently sent out parties of soldiers in search of elephants; to these men ivory took a secondary place to "meat," naturally, therefore, they cared very little for the ivory, and the results of these battues were frequently deplorable. I remember once witnessing one of these parties return with "meat" from two young female elephants and in the canoes they had also brought with them the dead bodies of two baby elephants which they had deliberately killed.

The two remaining products to-day are gum copal and

palm oil. In the closing year of the Congo State the former was certainly exploited *en regie*, but mainly in th'ose districts where exhaustion was overtaking the rubber forests. The latter produce has never formed any appreciable article of export.

Gum copal is to-day found in almost unlimited quantities in many parts of the Equatorial Zone and throughout the towns and villages the traveller meets natives everywhere engaged in its preparation. The gum taken from the upper part of the tree and near the surface of the earth is excellent in quality and much of it would easily command 1s. a lb. in Birmingham or London. The natives, however, readily accept 2d. per lb. but with any degree of competition prices would of course rise. Several companies are buying to-day faster than they can export.

Whilst passing through the towns, we were frequently assailed with the cry, " White man, won't you buy our copal ? " I questioned some of the merchants upon the possibility of an early exhaustion and was informed that in the Equatorial regions the exudation, if removed, replaced itself within a single season. My observation of some hundreds of copal trees in different areas leads me to regard this as a somewhat optimistic statement. It is certain that considerable profit can be made from the purchase and export of this virgin product, for at the rate now ruling it can be purchased and transported to Europe at an inclusive cost of about 4d. per pound.

Palm oil exists all over the Congo. In many districts the palm forests cover several square miles, but whether it can be produced at a profit is somewhat doubtful.

There remain, therefore, but two actual virgin products

GUM COPAL FOR SALE, UPPER CONGO.

GOVERNMENT IVORY AND RUBBER, UPPER CONGO.

possessing any certainty of a future—copal, and the fruit of the palm tree ; rubber can only be regarded as an ever decreasing asset.

What, then, are the potential assets ?

In the mineral world there are some possibilities in gold, diamonds and copper, but all these are somewhat doubtful assets and contribute but little to the general welfare of the community which must rest primarily upon agricultural development.

Almost any tropical product will grow in the Congo, for the area is so vast that it provides land suited in one part or another to coffee, cotton, rubber, cocoa, hemp and corn. The product of the future will not be determined only by the nature of the land upon which a given article can be grown, but rather by the one that is most suited to the native agriculturist.

The real difficulty is that few Belgians seem capable of thinking anything beyond rubber on the one hand, and the native as a servile labourer on the other. Colonial opinion in Belgium and on the Congo itself appears to be firmly wedded to this restricted view of colonial expansion. This circumscribed vision can comprehend the serf, the labourer, or the domestic slave, but the free, industrious and successful coloured citizen, carving out an economic future, in which the State can indirectly share, is apparently beyond the mental horizon of most of those who at present control the destinies of the Congo tribes. True statecraft would have placed a halo round Annexation Day, making it one of great rejoicing throughout the Congo by declaring that through the action of a generous Administration rubber collecting by the State would from that date cease for all time. But through lack of colonial imagination this great opportunity for regaining the confidence of the native tribes was thrown away, and the Administration rehoisted the old Congo State flag with a miniature Belgian flag relegated to the corner, at the same time letting it be known that upon rubber production—the synonym of horror to the native mind—the future would depend.

The failure of the rubber cultivation enterprise is complete. Whatever the man in the street may think, the Belgian Government knows that Monsieur Renkin's scheme for relieving the Belgian Exchequer has utterly failed. The twenty to thirty millions of *productive* rubber trees dangled before the eyes of the Belgian tax-payer exist only on paper.

Cotton has been proposed, but what possibilities has cotton cultivation, not only in the Congo but anywhere in West Africa, where it comes into competition with cocoa or palm oil? Cotton requires that the worker should toil under the fierce rays of a tropical sun; it demands constant attention if it is to be kept free from weeds and undergrowth, and when the harvest is gathered the native can never receive the financial reward which attaches to palm oil and kernels or to cocoa. In a crude way the West African is a careful mathematician, and though in his primitive condition he knows nothing about square yards, acres and compound interest, he can soon tell what products he can grow most profitably on a given piece of ground—and cotton is not one of them.

If the Belgian colonial authorities could divorce themselves from rubber and concentrate on cocoa they might yet turn the Congo wilderness into a garden. A few enterprising Belgians have already seen possibilities in the cocoa bean. Its cultivation is at present undertaken by the Belgian Government, the Roman Catholic Missions, and by a few small companies. The principal area is that of the Mayumbe, a compact territory between the Belgian Congo and the Portuguese river, the Chiloango ; there are other plantations a thousand miles from the mouth of the Congo on the banks of the Aruwimi and also of the main Congo, but these latter are characterized by such neglect that no one regards them seriously.

It is difficult to imagine a tract of country more ideally suited to the cultivation of cocoa than that of Mayumbe. The hills and valleys abound in water-courses, the soil is good and the climate reminds the traveller very much of the Gold Coast territories. Some of the plantations run for miles along winding valleys, but the great trouble with Mayumbe is that perpetual nightmare— common to the whole of West Africa—scarcity of labour !

Within three days' steam of the Congo, the British colony of the Gold Coast has solved the question of labour, has started an industry which gives the native producer a return of over a million and a half sterling per annum, has provided the European consumer with a great cocoa area which twenty-five years ago produced little beyond internecine warfare and jujus, and yet the Belgian Government has never even given a practical consideration to this unique example of colonial expansion which could so easily be applied to the Congo.

Rubber and cotton have but a small future in the Congo. Sisal, gold and copper have a possibility, but

cocoa, the products of the palm tree, and any other
vegetable oils, give promise of a real future, provided cheap
transport and sound statesmanship are forthcoming.

An oppressive sense of hopelessness affects the traveller
in the Congo as he speeds up and down those mighty
rivers, across the numerous lakes, or tramps through the
silent forests.   He sees the possibilities of that land, the
earth he treads gives forth an intoxicating odour of
fertility.   The tribes amongst whom he lives and moves
are nature's children and the little incidents of daily
travel impress him with the fact that, given a chance,
those sturdy bodies and stout limbs could turn Congo-
land into a paradise of affluence and luxury.   Then, as
he muses on these things and dreams of ideal homes and
villages, and tropical plantations pouring forth exchange
values of oil and cocoa for cotton goods and hardware,
the practical mind, like Newton's apple, comes down to
earth again and weighs actualities and asks the pertinent
question—" Can Belgium do it ? "

The Congo demands large financial aid from the
Mother country.   This is a fact which has never been
realized by the ordinary Belgian—and he might object
if he knew.   Even the British subject, whose colonial
conception has grown with him from childhood, has very
little idea of the large sums of money which are found
by Great Britain towards aiding her Crown colonies along
the path of progress.   Belgium cannot expect to run the
Congo successfully without large drafts from her home
Exchequer ;   her colony, measuring nearly a million
square miles, will require at the very least a million
pounds sterling per annum for twenty years.

Belgium can beyond question find that sum of money, providing her people are prepared to share the black man's burden which their late Sovereign made so heavy. The difficulty, however, is that King Leopold and his entourage made such prodigious fortunes that the Belgian people have always regarded the Congo as a veritable El Dorado. The Belgian colonial authorities reiterated again and again, until quite a recent date, that the Congo would never involve the nation in financial sacrifices. Couple this impression, so wickedly fostered by politicians who should have known better, with the fact that the Belgian has no colonial conception, and the reader will agree that any statesman will have a difficult task in persuading the Belgian nation to make large and continuous grants from the Home Exchequer.

The British conception rests upon a profound belief in the old scriptural paradox : " He that loseth his life will save it." The Colonial Office in Downing Street does not, like its sister bureau—the Foreign Office—display texts of scripture on its ceilings, and the Colonial Secretaries might not in this material age admit scriptural guidance in Imperial affairs, but woven into the fibre of our administration is a basis of Christian philosophy which, though it admits occasional incidents of a regretable nature, yet pursues in the long run the straight course of sacrificing men and money for backward nations and countries, quite regardless of consequences. The cynic will say, " Yes, with the certainty that the goose well cared for will lay golden eggs." Certainly, but that is part of the Divine contract for pursuing that which is right. This, however, is what few Belgians understand

—or any other colonial Power for that matter—but it is part and parcel of colonial statecraft without which tropical colonies at least can never be a success.

The financial problem, difficult though it may be, is the easiest of solution. That of finding the men is at present insoluble. This is, in part at least, due to another fatal error made by Belgium when she annexed the Congo—the retention in her service of all the old Congo officials. They are there to-day, many of them pressing on to higher positions in the colony. The fact that these men, trained to oppression by King Leopold and openly upholding the old Leopoldian conceptions, are still in high favour does not escape the quick-witted native, and of course tends to alienate still further the native and governing communities.

There are, however, other dangers arising from this situation. These " old hands " are educating the juniors, and in the process are instilling into their young and inexperienced minds a dissatisfaction with present conditions and emphasizing to them that the older system of " teaching the natives the dignity of labour " was better all round. They are always careful to add " without atrocities, of course," but what they cannot see is that the old Leopoldian system was impossible " without atrocities." It will be readily agreed that when the burden of the Congo begins to make itself felt upon the Belgian nation these reactionaries—" Men from the spot," " Men of long experience "—will find a ready echo throughout Belgium. Again, as in the financial position so also in the administrative future of the colony, the call comes for the really bold statesman, strong enough to

break completely with the past and to clean out of the
Congo these *soi disant* administrators, who, incapable
of appreciating colonial requirements, should return to
their original employments of running music halls, tram
driving, breaking stones on the highway, 'bus conductors,
waiters, bricklayers, clerks, and so forth.

" How," I am often asked, " could these men be
replaced ? " First, the very fact that such men are no
longer in the service would undoubtedly attract the better
families of Belgium, for it may be remarked that many
of the merchant houses are able to obtain an excellent
type of man. I asked some of them why they did not
enter the Government service, but almost invariably I
received this kind of answer : " What ! join a service with
A——in it ! " " What ! accept a position under B——! "
These replies were eloquent and convincing to one who
easily realized how utterly impossible it would be for the
better type of man to associate with " A—— " and
" B——," their records being so well known in the Congo,
however much they might be covered up at home. Here
again is further evidence of the lack of colonial imagination amongst the higher officials in Brussels. If Belgium
cannot find—as admittedly she cannot—a sufficiency of
experienced men in Belgium, cannot she find them in
France and England ? She can find them, of course, in
both countries, but hesitates to employ other nationalities
for the higher positions, with the result that very few men
are prepared to accept positions with futures " only for
Belgians."

A Scandinavian captain recently gave me a good
example of the results of this folly. He informed me that

a friend of his reached Stanley Pool one day with his ship after an up-river journey of three weeks. Arriving at " The Pool," as the upper river port is designated, the then superintendent of the marine—who, it was openly stated, knew more about the manufacture of cheap pickles than stevedoring—instructed him to load up 90 tons of cargo and sail within three hours !

In vain the captain protested that it could not be done in the time, and the only reply he received was a batch of natives hurried down to bundle the cargo pell mell on board ; they pitched the cargo into the holds in any order and the captain heaved up his anchor and got away as instructed " within three hours," but the task of sorting the whole cargo at every little post over that 1000 miles' run, turned a normal journey of two weeks into one of over a month. I cannot vouch for this incident, but it is typically Congolese.

The Congo territories denuded of their stores of virgin wealth, with no new sources in sight ; the people decimated and disheartened ; the Home Government possessing no Colonial experience, and still worse no Colonial conception ; the local officials still firmly wedded to the old theories, constitute anything but a happy augury for the future. That Belgium possesses many men animated by the loftiest sentiments is beyond question, but mere sentiment does not meet a situation which requires a broad outlook, a large experience and real sacrifice both in men and money.

PART IV

MORAL AND MATERIAL PROGRESS

I

THE PRODUCTS OF THE OIL PALM

WITH the date palm we have been long familiar, the cocoa-nut palm likewise, and those too which decorate our ball-rooms, galleries and banqueting halls, we greet as delightsome friends, but what is the oil palm—the Eloesis Guineensis of West Africa ? It is said that five thousand years ago its sap was used by the Egyptians for the purpose of embalming the bodies of their great dead. To-day by its aid we travel thousands of miles at express rate ; it has been so handled by modern science that it enters largely into our diet ; the merchants in Hamburg and Liverpool make fortunes out of it ; millions of coloured people live by it, and yet it is barely known to the civilized community. A fortnight's fairly pleasant steam from Liverpool brings the traveller in sight of the high red clay coast line of Sierra Leone, and there the oil palm first greets the traveller in all its luxuriant grandeur.

From Freetown away down the coast as far as San Paul de Loanda, the traveller is never far from the home of the oil palm—the most valuable tree of West Africa—probably the most prolific source of human sustenance in the world. She greets the traveller everywhere as he steps ashore ; she invites him to the cool shade of her avenues leading to some hospitable bungalow ; she affords a shelter at intervals along the scorching dusty track—as welcome as an oasis of the desert ; she waves at him vigorously from the hill-top like some fluttering banner, or gently nods her graceful plumes in the still valley ; she stands as sentinel on the outskirts of the native village, or like some giant memorial column on the plain. All nature strikes the African traveller dumb with admiration, but above all in entrancing loveliness the graceful oil palm reigns supreme.

To the parched and weary she is at once meat and drink and friendly shelter. Her palm cabbage and nut oil are no less palatable than her foaming fresh-drawn wine, and if no other home affords, her branches offer a temporary and not comfortless dwelling. She provides her guest with oil to lubricate his gun, with fibre to plug his boat if it springs a leak ; her fronds serve as a weapon to combat the infinite torment of flies, or interlaced as a basket to carry a meal. To her the native goes for a tool or a cooking-utensil, a mat or a loin cloth, a basket or a brush, a fishing net or a rope, a torch or a musical instrument, a roof or a wall. To him she is a necessity, to the traveller a luxury, to the merchant a fortune, to the artist a subject full of charm.

Professor Wyndham Dunstan has stated that the oil palm " does not occur thickly much beyond 200 miles from the coast." Since those words were written we have learnt that whole forests of the oil palm exist over a thousand miles from the coast. It thrives throughout West Africa wherever the atmosphere is sufficiently humid, but it loves best of all the swampy valleys of Sherboro sland and Nigeria, the cocoa farms of San Thomé, the

AN AVENUE OF OIL PALMS. 10 YEARS' GROWTH.

Gold Coast and the Congo, which are by it provided with the necessary protection from the scorching sun and from the fierce tornados which sweep periodically over the land. In the strictest sense the oil palm has never yet been an object of cultivation in West Africa, neither is it in the literal sense self-propagating. The housewife, separating the fibrous pericarp from the nuts, tosses the latter aside or scatters the residue on to the rubbish heap behind the hut, with the inevitable result of an early and vigorous crop of young palms. In the course of time the inhabitants of the village, according to African custom, pick up not only their beds, but also their huts, and walk, perhaps something less than a mile away, where they clear another piece of forest land and build up another village. The old site, thus abandoned to nature, is quickly covered with vigorous growth, but in the race for supremacy the graceful palms lead the way and become the communal property of the former inhabitants.

The screeching grey parrot of West Africa with its horny bill tears the oily fruit from the bunch, after consuming most of the oleaginous fibres of the pericarp, drops the nuts whilst flying, far and wide. These, in turn, add to Africa's economic wealth, and thus do man and animal join in spreading through ever wider regions the growth of the oil palm.

Within ten years the tree begins to push out its bunches of fruit, beginning with tiny bunches of the size, shape and appearance of an ordinary bunch of black grapes. Some trees bear in eight years, and an earlier date still is claimed for certain varieties, but the fruit at this stage seldom yields any appreciable quantity of oil. From fifteen onwards to a hundred and twenty years, the palm plantations give forth an almost continuous supply of fruit, every tree bearing twice a year. In the rainy season the supply is most abundant, but in the second period, known by many as the " short wet " season there is a fair secondary harvest. All the trees do not, however, bear at the same time, and in many areas of the Equatorial regions where the seasons are not sharply defined or always regular, the supply of nuts is never exhausted.

In appearance a head of fruit resembles a huge bunch of grapes with long protecting thorns protruding between each nut, and a good bunch will contain from 1500 to 2000 nuts. A single fruit in appearance is about the size of a large date, and the pericarp is composed of fibre matted closely together with a yellow solidified oil, which fibrous substance envelops a nut or " stone " ; this in turn encloses a kernel of the size and shape of a large hazel kernel, in appearance and composition indistinguishable from the well-known " Brazil nut." From the fibre a dark reddish oil is obtained, whilst the kernels that are shipped to Europe yield a finer white oil.

The almost universal practice amongst the natives in harvesting the nuts is to climb the tree by walking up the trunk with the aid of a loop of stout creeper. Arriving at the top at a height of sixty or eighty feet, the man deals a few vigorous blows with an axe which severs the bunch or bunches from the tree and they then fall to the ground. As the whole family usually takes part in the production of oil and in the division of labour, the man, having cut down the fruit, descends the tree, picks up his protective spear or gun, and returns home, closely followed by the wife and daughters, who transport the bunches of nuts in the wicker baskets which they have woven in their spare moments.

In every colony a similar process is adopted to separate the fruit from the parent stem. Until it is over ripe, the fruit not only adheres firmly to its stem but the porcupine thorns sometimes two inches long, make separation anything but a pleasant task. The tribes everywhere collect the clusters or bunches into heaps and cover them with plantain or banana leaves, exposing them to the sun for from three to six days, the effect of which is that the nuts, subjected to the hot rays of a tropical sun and cut off from the refreshing sustenance of the mother tree, lose their tenacious grip and readily drop away from their stem.

The methods adopted to force the oil from the fibrous pericarp differ considerably in the several political divisions of West Africa. Roughly, however, they fall into two divisions : (a) by fermentation ; (b) by boiling ; and in certain parts of the Kroo Coast by a combination of both methods.

The fermenting process is carried out by placing a large quantity of separated, but hard, nuts into a hole about four feet deep, this having been first lined with plantain leaves. In the regions nearer the coast towns, these pits are either paved or cemented inside and in some cases they are both paved and cemented. The nuts are covered up and then left for some weeks, even months, to ferment thoroughly. They are then either pounded in the pit with wooden pestles, or they may be taken out and treated in prepared wooden mortars.

The process of boiling is more expeditious. The nuts are boiled or steamed until the firmly coagulated fibre shows signs of yielding ; then they are placed in an old canoe or large mortar and pounded with wooden pestles. In both processes, whether by fermentation or by boiling, the oily fibre separates itself from the hard inner " stone." The fibre, which is by this time a tangled mass of yellow and brown, is then taken and squeezed, sometimes with the aid of water, through a woven press and a stream of golden liquid results. Sometimes loads of the oily fibre are thrown pell-mell into a large canoe half filled with water in which the children delight to paddle, causing the oil to rise to the surface, when the elders skim it from the top and carry it in earthenware pots for boiling and straining before sending it on its way to the market and the European consumer.

The oil, however, is but one exportable product of the palm tree ; the value of the inner kernel may be gathered from the fact that over four million pounds' worth of palm kernels are sent to Europe every year. This kernel is encased in an extremely hard shell, which varies so much in size that until quite recently there was no satisfactory " stone " cracking machinery in Africa. There are now several machines on the market, but the old grey-haired lady of the West African kraal, with her primitive upper and nether grind stones, still makes by far the most reliable " cracker."

At Victoria, in German Cameroons, we saw an elaborate set of machinery for dealing in turn with the

"WALKING" UP TO GATHER FRUIT. WEAVER BIRDS'
NESTS ON THE PALM FRONDS.

HEADS OF OIL PALM FRUIT.

oily fibrous pericarp of the nut, and later, extracting the kernel from the inner stone. The latter process was that of a general crushing, then throwing the entire mass into a brine bath and so separating the shells from the kernels, which were then taken out and dried in the sun. This process, while being infinitely more expeditious, has the obvious drawback that a large proportion of the kernels are so bruised and broken that it entails a considerable wastage of oil.

THE PALM IN TONNAGE AND IN FIGURES STERLING.

Exports in round figures for the year 1911—

	Oil.		Kernels.	
	Tons.	Values.	Tons.	Values.
French Senegal . .	1	36	1,418	14,300
„ Guinea .	53	1,300	4,500	36,600
„ Ivory Coast .	5,800	107,100	5,340	45,500
„ Dahomey .	14,400	254,100	34,200	400,000
„ Congo . .	125	3,100	570	7,600
British Gambia . .	—	—	447	4,758
„ Sierra Leone .	2,902	69,930	42,893	649,347
„ Gold Coast .	6,441	128,916	13,254	175,891
„ Nigeria . .	77,180	1,696,875	176,390	2,574,405
German Cameroons .	3,000	63,000	13,500	177,530
„ Togoland .	3,050	61,600	8,100	101,700
Belgian Congo (approximately) .	700	20,000	2,500	40,000
	113,652	£2,405,957	303,112	£4,227,631

TOTAL OUTPUT.

	Tons.	Values.
Oil . . .	113,652	£2,405,957
Kernels . .	303,112	£4,227,631
	416,764	£6,633,588

The proportionate output from the palm trees from the different colonies of West Africa is therefore—

	Square mileage of territories.	Tons.	Values.
French . . .	992,000	66,407	£869,636
Belgian . . .	900,000	3,200	60,000
British . . .	454,160	319,507	5,300,122
German , . .	224,830	27,650	403,830

Production in figures sterling per square mile under the several colonizing Powers—

	£	s.	d.	
Great Britain	11	13	3	per square mile
Germany	1	16	0	,, ,,
France	0	17	8	,, ,,
Belgium	0	1	4	,, ,,

Whilst the palm provides one of the principal exports of West Africa for consumption in Europe, its domestic uses are inseparable from native life. The fruit is used by the natives in many sections of primitive culinary art. Pounded with manioca leaves, Indian corn and red peppers, a savoury pottage is manufactured which is a universal delight. One of the choicest vegetables in the African continent is the pearly white head of the palm, which, in small trees of two years' growth, weighs about 1 pound, but in trees of many years' growth, may turn the scale at 56 pounds. The substance of this vegetable differs in appearance and taste but little from the Brazil nut, but when cooked provides a succulent dish not unlike, though superior to, sea kale. The natives cook this in palm oil, but Europeans usually prefer it boiled and served with a white sauce, or baked in a custard. To obtain this vegetable is almost invariably

The Creeper at an early stage.

Root and Branch in deadly grip.

THE OIL PALM IN THE GRIP OF ITS PARASITIC ENEMY.

to destroy the tree, consequently it seldom figures on the every-day menu.

Meat, fish and fowl are all of them stewed in palm oil, and, as African meat is deficient in fat, the palm oil makes an excellent and appetizing substitute. I once smelled a very savoury native " hot pot " which, upon examination, revealed a wonderful mixture. The liquid was golden with palm oil, and floating about, adding to the compendium of flavours, I detected bats and beetles, a flat fish "cheek by jowl" with a monkey's head, caterpillars fitting themselves in with sections of field rats and parrots—altogether a stew delightful to the nostrils, at least of the African boys and girls who squatted around the huge clay cooking pot. The white man, though he usually has no keen appetite for native stews or pottage, lunches and dines off " palm oil chop " with as great a relish as does his Indian confrère upon " curries." The " chop " may be fish, flesh or fowl, but it all goes by the name " palm oil chop," which has a happy and almost essential knack in West Africa of hiding a multitude of " foreign bodies."

No African meal can be regarded as complete without the addition of palm oil, and, as a beverage, palm wine is extensively though moderately consumed. This sparkling beverage closely resembles in appearance the " stone ginger " of civilization. The tribes on the Upper Kasai are probably the greatest consumers of palm wine in Africa. In those parts of the tropics where quantities of sugar cane are cultivated, palm wine competes with a sister product from the cane ; the sweet and somewhat insipid taste of the latter being more palatable to some tribes than the sharp flavour of the palm wine. The Eloeis wine is the sap of the palm tree itself, extracted by various means, generally by cutting off the male flower-spike and fixing a calabash to the wound to catch the juice which is removed every morning. Another method is to remove the palm cabbage or head ; yet another, to cut down the tree and " dig " a hole in the heart of the trunk, from which the liquid is then scooped into a calabash or earthenware pot. Europeans generally prefer the wine when fresh from the tree, owing to the fact that after a few hours it begins to ferment and loses its sweetness.

The oil palms of West Africa are taking an increasing share in supplying the temporal wants of both the white and the coloured man. It is safe to say that there is no tree in the universe capable of providing to so great and varied an extent, the daily wants of the human organism.

FINE HEADS OF OIL PALM FRUIT.

II

THE PRODUCTION OF RUBBER

RUBBER has been known for the last four hundred years, but it is only within the last century, or little more, that it has been put to practical use. Civilization was for nearly three hundred years content with the historical fact of Pincon's Indians of Brazil playing "ball" with crude lumps of rubber, and then it awoke to the fact that rubber could be used to erase pencil marks. In our boyhood Charles Macintosh had established its use as a protective from rain, but in our manhood the annual demand of Great Britain alone for rubber has grown to nearly 50,000 tons. We have lived through the sensation of a "Rubber Boom" which is only now commencing to exact its toll for the immeasurable folly of the thoughtless investing public.

The native use of rubber in West Africa as also among the Brazilian Indians, was first as an aid to merrymaking, in the form of heads of drum-sticks, and in that capacity evoked harmonious chords from the goat-skins tightly stretched over the hollowed forest log. How little these early Africans dreamed that this simple aid to the charms of music would one day deluge their Continent in human blood! There are to-day very few colonies in West Africa without rubber forests which nature—prodigal here as everywhere with her economic gifts—planted generations ago.

The discovery of the great West African rubber supplies dates back about thirty years, but it is a remarkable fact that Stanley in his books on the Founding of the Congo Free State, laid very little stress upon the future of rubber in the Congo.

In 1882 Sir Alfred Maloney urged Southern Nigeria to wake up to the possibilities of rubber, and in 1894 Sir Gilbert Carter, to whom our Nigeria colony owes so much, invited a party of Gold Coasters to explore the hinterland forests with the result that they discovered an abundance of what appeared to be rubber-bearing plants and trees. The native community then set about vigorously searching for rubber with the result that the " Ireh " tree was discovered, and specimens of its latex forwarded to Kew in 1895. Although it had been discovered in the Gold Coast colony ten years earlier the administration in Nigeria was apparently in ignorance of the fact.

There is some evidence that King Leopold received the first intimation of the almost fabulous stores of rubber in the Congo forests between the years 1888 and 1890, and the alert mind of that astute monarch lost no time in formulating plans for its exploitation in the Congo Free State, and what is less generally recognized in the French Congo also.

Since 1885, when the African product first made its presence felt in the rubber market, the natives of that continent have gathered and sent to Europe over 250,000 tons of rubber, the outstanding fact being that all this latex represents sylvan produce, the replacement of which is extremely doubtful. Dr. Chevalier is of the opinion that the natives themselves received for the total output 500,000,000 francs or approximately 9*d*. per pound. This I very much doubt, for it must not be forgotten that a large proportion of the rubber was obtained, if not for nothing, then for very little.

The principal sources of rubber latex are the Funtumia (Ireh) and the Landolphia varieties, which, to the ordinary reader, fall respectively under the classification of trees and vines. The full-grown Funtumia tree measures from 2 ft. 6 in. to 4 ft., or more, in circumference. The growth of the Landolphia is wild and erratic, creeping along the ground sometimes for several yards, then gradually winding its way through the undergrowth and away up the limbs and branches of the firmly rooted forest giants to a height of forty to fifty feet, then in the full enjoyment of light, it becomes vigorously prolific, sending its leafy branches in all directions, and interlacing the trees overhead. Most scientists seem to agree that it is only when the Landolphia emerges into the sunlight at the tree tops that material size is imparted to the main stem. From the economic standpoint it is important to bear in mind that ordinarily a Landolphia vine takes from ten to twenty years to climb its way up the tree trunk of the average forest tree, at which period the main stem of the vine is seldom more than one inch in diameter.

Beyond question by far the larger proportions of rubber from Central Africa have been obtained from the Landolphia vines, that from the Congo basin almost entirely so. Next in order comes the output from

Funtumia forests of the more northerly latitudes, and beyond this a certain amount of grass rubber has been obtained, but the results barely justify the trouble involved.

The extraordinary development and almost general investment in the rubber industry have familiarized the public with rubber production. Almost every schoolboy could write an essay upon the herring-bone or half herring-bone tapping, coolie lines, spacing and so forth. The production of rubber conveys to most minds well-ordered estates of upright trees, model workmen's dwellings, drying and boiling sheds, constructed by skilled Europeans, rolling tables, hot and cold water supplies, all under the control of neatly clad coolies. None of these conditions apply to West Africa, for there everything is to-day primitive.

The larger Funtumia trees are tapped in a very rough " herring-bone " manner and the latex caught either in leaves or in a calabash, and then transferred to a wooden receptacle for coagulation, but large numbers of trees have been bled to death through the almost incessant tapping to which they have been subjected. Funtumia more than any other variety requires carefully-regulated tapping, and it is well-nigh hopeless to expect the native collector in the hinterland regions to exercise that degree of care which the Funtumia tree demands as the price of giving forth a sustained output. The damage done to the bark alone in the rough and ready methods of extraction almost invariably renders the tree unfit for future tapping ; the trees will live sometimes for a few years, but before long they perish.

Dr. Chevalier, writing of the Ivory Coast, says : " Wherever exploitation has spread it has caused the adult Funtumia trees to disappear very rapidly. Some are cut level with the ground by the natives in order to extract their maximum yield, others, tapped too frequently, die standing, at last there remain only young Funtumia trees, under fifteen years of age." This is true of the major part of the rubber-bearing regions of West Africa.

Several methods are followed in the extraction of the latex from the Landolphia. In every case that has come under our notice the vines were cut down with little thought for the future. Indeed in the upper regions of the Congo the natives sever the vine close to the ground and then tearing it from the trees to which it clings, they cut the vines into lengths of about eighteen inches and pile them into stacks so that from the severed ends the latex may bleed into forest leaves or gourds. Many of the tribes raise the stack of severed creepers upon forked sticks and kindle a slow fire beneath as they assert that the latex flows more freely and completely with the application of heat.

The whole process is beyond question most wasteful, particularly where the natives not only sever the vine, but dig up the roots, compelling these also to yield up their stores of latex. To-day as the traveller marches through the rubber forests of the Congo basin he meets every few yards little heaps of decaying vine from which the rubber has been taken. Frequently too, one sees overhead a tangled mass of dead vine which has withered away through the main stem having been severed. The natives were either in too great a hurry, or else unable to climb for those spreading vines which would often measure some hundreds of yards.

Another method is that adopted by the native tribes in the Kasai River of the Congo, and the Lunda province of Portuguese Angola. Whole families or tribes will make a temporary home in the forest, pitching their little huts on a piece of high ground near a stream. Every day the men will scatter in all directions cutting down and gathering the vines into bundles which they will convey to these little encampments.

The bark of the vines is then stripped off and laid out on blocks of wood, old canoes, boards, or trunks of trees, preparatory to beating it with heavy wooden mallets, which process gradually reduces the bark to a stringy mass not unlike shredded tobacco. It is then threshed with smaller mallets which in time gradually pulverize the wood element into fine powder, leaving " pancakes " of red rubber, about the size of a breakfast plate. These are then cut into thin strips, starting from the outer edge, and wound into balls, just as the manufacturers wind balls of knitting wool. This method though equally wasteful in collection, conserves the whole of the rubber latex.

Travellers in the Kasai territories of the Congo are generally first aware of their approach to human habita-tion by hearing the distant thud, thud, of the rubber mallets which is a feature of almost every village of that region.

Hand in hand with the rubber work of the Congo is that of cane basket making, which the busy women weave

CARRYING RUBBER VINES TO VILLAGE.

EXTRACTING RUBBER, KASAI RIVER, UPPER CONGO.

in all sizes for packing the rubber, thus avoiding the heavy cost of importing " shooks " or barrels from Europe. Every year some hundreds of thousands of these light but very strong hampers are made for conveying the rubber to the buying stations and thence to the European markets.

The West African rubber problems of to-day which overshadow all others are those of exhaustion and replenishment. Are the forests denuded of rubber, and if so, is there any probability or possibility, of rubber cultivation to replace the exhausted supply ? Both these phases of the question are difficult of complete and categorical answer.

For thirty years now exploitation has been running wild through the forests, and within the last fifteen years the rate and methods of exploitation have from every point of view been ruinous. The Funtumia trees have been ruthlessly cut down and even where tapping has taken place, it has been done at any and every season of the year, and in general practice tapped whenever and wherever the tree would yield an ounce of rubber.

Dr. Chevalier is of the opinion that the Funtumia will replace itself owing to the remarkable habit of self-propagation which the tree possesses. The light feathery seeds are easily carried upon every breeze it is true, but unfortunately there is little hope of preserving these young trees from crude and reckless tapping in the farther recesses of the forests. It is generally accepted that the rubber vine areas are being rapidly exhausted. Mr. Consul Mackie says of the Congo, " Wild rubber in districts in which it has been worked on an extensive scale, is now becoming scarce in places. Many of the large rubber zones have been worked out completely."

We were informed by natives of the Kasai who were bringing in their rubber to the factories, that whereas ten years ago they had only to go one or two days into the forests before finding rubber, they now have to journey nearly a fortnight before they can locate any appreciable number of vines. Throughout the Equatorial regions of the Congo, the rubber vines and trees are so completely worked out that the natives have given up attempting to collect rubber and devote all their energies to gum copal and palm oil.

Most disinterested " coasters " will support Dr. Christy in the opinion that if the African rubber industry is to depend upon the wild forests there is very little chance of its survival.

Within the last fifteen years efforts have been made in various colonies to cultivate rubber. The most promising results are certainly in Nigeria, where the Benin communal plantations are proving so successful that villages in other districts are commencing similar plantations. Many thousands of Funtumia trees are now ready for tapping and some of the rubber obtained has secured 6s. 6d. per pound. Individual native farmers are now taking up rubber planting, and in Southern Nigeria we saw some well-ordered plantations under native control, one of which started in 1896 has over 30,000 trees and gives promise of a good output. In the Gold Coast the natives are interspersing Funtumia trees with their cocoa plants, under the instruction of Government advisers.

In Belgian Congo vigorous efforts have been made for the last twelve years to cultivate rubber. In the year 1899 a Royal decree was issued requiring that 150 trees or vines should be planted for every ton of rubber exported, and in June, 1902, the number of plants was raised to 500. As a further incentive some of the Concessionnaire Companies gave a bonus to their agents for every tree planted. The ordinary Belgian being very keen on piling up his banking account the planting was pursued with vigour. As, however, the ordinance did not specify the variety to be planted the Agents of the State and Concessionnaire Companies planted varieties good and bad, known and unknown! until on paper the total number of trees planted ran into many millions.

Every few months an Inspector was supposed to visit these areas, but as this official usually had an area of about 25,000 square miles under his control, he was seldom able to visit more than one centre every year. Badly paid, with little allowance for provisions, this man usually responded to the warm hospitality of his planter host, and generally did not make exhaustive inquiries into the rubber planting. On one occasion such an inspector visited a district after the Agent had gone to Europe, in order to " check " the trees and vines before the new Agent arrived to take over the stock and plantations. He asked me if I could direct him to one plantation of 60,000 trees and vines of which he possessed a neatly drawn chart. I could only direct him to where the plantation was supposed to exist, and he immediately set off on what I hinted was a useless journey, and as I expected returned in the afternoon without having discovered a single vine !

Apart from these paper plantations there are certainly several millions of rubber trees in the Congo, and every species almost has been tried. At one time the Belgian taxpayer was told that the Manihot Glaziovii was going to provide fabulous returns, but when the floods came and the winds blew, the spreading Manihots caught the force of the elements and toppled over in all directions like ninepins. The Funtumia was then going to save the Congo from financial disaster, but the " borers " took a fancy to the tree and this, coupled with the fact that in the Congo the Funtumia yields but little rubber, all serious attempts at the extension of Funtumia have been abandoned.

Hopes are now being centred upon the Hevea Brazilensis, but though many of these trees are of ten years' growth the yield is equally disappointing.

In German Cameroons rubber planting is being pushed forward mainly with the Funtumia and Hevea varieties. In Portuguese West Africa hopes are centred upon Manihot and Funtumia.

The best that can be said of the rubber cultivation in West Africa is that it has not yet passed the experimental stage, and that there is some promise of success in the Gold Coast and Southern Nigeria.

There is, however, one other factor which must not be overlooked, Mr. Herbert Wright pointed out last year that cultivated plantation rubber would soon be arriving in quantities which would cause embarrassment to the rubber merchants. It is certain that when this happens prices are bound to fall, perhaps dramatically. The question for the West African rubber planting community to ask is : can they, when prices fall, compete with the West and East Indies, where labour is plentiful and cheap, and where there is practically no costly land transport. A merchant from the Straits Settlements once informed me that West African rubber producers must be prepared to compete with the East—at 9d. per pound. If that prediction should be justified by future events then West Africa will be wise to concentrate upon its trusty friends the Oil Palm and Cocoa Tree.

II

THE PRODUCTION OF COCOA

Cocoa to most individuals is suggestive of carefully and tastefully packed tins, or in chocolate form, of delightful little packages done up in neat silver paper and prettily tied with bows of silk ribbon. To others it means a welcome and fragrant breakfast or supper beverage. To few, indeed, does it represent anything else. The man in the street, if he thinks at all upon investing his savings in cocoa, argues that after all there is a limit to human digestion, particularly where sweetmeats are concerned, consequently he need not trouble himself about " futures " in cocoa for the field is at best a restricted one. It never occurs to him that the demand for every species of vegetable oil and fat is becoming more clamant every day. Somehow he never asks himself why Bournville, York and Bristol cocoa is 2s. 6d. per pound, and Dutch only 1s. per pound. He presumes, and if he tries it he knows, that one quality is better than the other, but it does not occur to him that there is something in the one beverage which is lacking in the other. The butter from the latter—the pure fat too expensive to eat, but not too expensive to incorporate in pomades for personal adornment, has been extracted.

COCOA ON SAN THOMÉ.   TERMITE TRACK VISIBLE ON THE TRUNK OF TREE.

There is no more rigid limit to the demand for cocoa than to the demand for rubber ; not only so, but nothing has yet appeared even on the horizon of our imagination that can take the place occupied to-day by the cocoa bean, both for internal and external consumption. This cannot be said with regard to rubber, wool or silk.

The total world's supply is to-day close on a quarter of a million tons of cocoa per annum.   The East and West Indies and the great Amazonian Valleys, have for generations poured their supplies into Europe, but it is only within the last thirty years that West Africa has made her influence felt upon the cocoa markets of Europe.   It is very difficult to obtain reliable evidence as to the colonists who first introduced cocoa to West Africa, probably the credit for it belongs to the Portuguese, whose love of colonization is everywhere evinced by the plants, fruits and grain which they conveyed in past years from one continent to another.

Given a humid atmosphere, a well-watered land and a tropical sun, cocoa will grow almost anywhere up to a height of nearly 1500 feet.   Of such lands enjoying atmospheric conditions highly suitable to the production of cocoa, there are nearly one million square miles in the tropical regions of the West African continent. San Thomé and Principe, with less than 300 square miles under cultivation, supply to the world's markets over 30,000 tons of cocoa every year ; if, therefore, but one quarter of the potential cocoa producing areas of West Africa could be brought under cultivation at the same rate, there could be produced over 25,000,000 tons of cocoa.

To-day cocoa is being cultivated in the German colonies of Togoland and Cameroons ; in the Portuguese colonies of Cabenda, San Thomé and Principe ; in the Belgian Congo ; in the Spanish island of Fernando Po, and in the British colonies of the Gold Coast and Nigeria. In all these the production has distinctive features.

From the standpoint of plantation arrangements and the application of scientific methods, the Portuguese in San Thomé are easily first. This no doubt is due to the fact that for over twenty years the planters have been concentrating all their efforts upon the cocoa bean. Throughout their whole area San Thomé and the sister island of Principe are under cocoa cultivation and the traveller never gets away from the sour odour of fermenting cocoa. A series of high hills and deep valleys with numerous rivulets represent the physical features of the islands. The hill ranges, for the most part, rise tier above tier, until they culminate in the Pico da San Thomé with an altitude of just over 7000 feet. The summit of the peak is seldom seen, for the island lies bathed in mists, which warmed by a tropical sun provide the ideal cocoa-growing climate.

The streams which flow unceasingly down the hill-sides are scientifically trenched so that a continuous supply of water traverses the cocoa groves all over the islands, and the farms in the centre of each group of plantations all enjoy a plentiful supply of excellent water.

The fermenting sheds are all of them organized in an up-to-date manner for which a knowledge of industries in other Portuguese colonies hardly prepares the traveller.

Nowhere throughout West Africa are there such scientific and elaborate cocoa drying grounds as one sees on these Portuguese islands. The majority of cocoa planters in West Africa are satisfied with cemented drying grounds in open courtyards. The cocoa is spread out to dry and left in the open not only during the whole day, but throughout the night. On several roças on the cocoa islands, the Portuguese have, at enormous expense, fitted up drying grounds which are mechanically moved into shelter whenever a storm threatens. Doubtless it is due to the great care exercised by the Portuguese in the work of fermenting and drying that their cocoa is so uniformly good.

Altogether there are nearly 300 roças on the two islands and, with one or two exceptions, they are in Portuguese hands ; there is a Belgian plantation, and one or two are owned by natives whose ability to make cocoa production a financial success is demonstrated by the fact that one who died recently left £6000 for the education of the children of San Thomé.

The cocoa plantations on these islands are all so compact and within such easy reach of the sea-shore that transport is quite easy. Both horses and mules live fairly well on the islands, and these, coupled with bullock carts and some 1500 kilometres of Decauville railway throughout the islands and running out to the pier, render unnecessary the porterage which constitutes such a problem for the cocoa planters in every other colony in West Africa. It is a melancholy thought that this industry, built up at so great cost to human life—both white and coloured—stands only a bare chance of permanence. The lack of indigenous labour, coupled with the absence of statesmanship on the part of the Home Government, can only lead to irretrievable disaster.

The nearest approach to the Portuguese systems of cocoa production is to be found in Belgian Congo, where physical and climatic conditions are almost identical with those of the Portuguese islands. The first plantations are met with close to the mouth of the river in the Mayumbe country, but before reaching the next one has to traverse nearly a thousand miles. These are situated at the confluence of the Aruwimi river and the main Congo, and there are besides several small plantations on the Aruwimi itself. As cocoa-producing enterprises, the only ones to take into serious consideration are those of the Mayumbe country, south of the Chiloanga—the Portuguese river, which enters the sea at Landana. The plantations are run under three separate interests, and may be classified as State controlled, Roman Catholic and Merchant. The merchants complain that their difficulty in obtaining labour is greatly increased owing to the missions and the State using forced labour for their plantations. It seems incredible that this should be so, but these complaints are neither new nor isolated. The Commission of Enquiry sent to the Congo by King Leopold had evidence before it which shewed that the Mission farms at least were largely staffed with forced labour. The following passage is an extract from the report of that Commission in 1905 :

" The greater part of the natives which people
" the chapel farms are neither orphans nor workmen

" engaged by contract. They are demanded of
" the Chiefs, who dare not refuse ; and only force,
" more or less disguised, enables then to be retained."

If the Belgian Government could concentrate upon a
serious development of the Mayumbe country by laying
down railways, making roads, building bridges, opening
up creeks, and rivers, there is no reason why the Mayumbe
country should not increase its yearly output of cocoa
by many thousands of tons.

The Spanish contribution to the world's supply is not
yet or ever likely to be anything material, for as colonists
in Africa the Spaniards have ceased to count.

In German colonies cocoa growing is extending
rapidly and from a financial point of view satisfactorily.
The German Administration in the Cameroons, however,
seems to favour such enterprises mainly as European
undertakings in which the natives are mere labourers.
Within recent years, probably in view of the success of
the Gold Coast production, some effort has been made
to encourage the natives by gifts of seed and young
plants to lay down their own plantations. But the
prevailing German opinion has been set forth in a German
report, published in *Der Tropenpflanzer* (No. 1, January,
1912), wherein it is stated :—

" What is required in the Cameroons is a more
" liberal policy on the part of the German Govern-
" ment towards the plantations, both as regards the
" terms for acquiring land, and on the part of the
" district officials to obtain better facilities for
" getting labour, in order to warrant and make
" possible a large and profitable extension of the

" cocoa-planting area. This will mean a material
" improvement in the prosperity of the colony, for
" it is evident that what the Gold Coast has achieved
" by means of an intelligent population, and under
" suitable climatic conditions, can and will never
" be done in Cameroons with such material as the
" Bakwiris, Dualas, etc."

Whilst colonial Germans take this view, the native certainly will never emulate the Gold Coast tribes, for the African has a habit of acting up, or down, to European expectations. The Editor of *Tropical Life* truly remarked that whilst these views are held in Berlin, " Germany would never do any good with the Bakwiris and Dualas ; neither did she with the Herreros, and so ' punished ' them because they, poor wretches, could not understand the German method of ruling Africa as do the German Michels at home."

There is some reason to believe that the cultivation of the cocoa bean began in Cameroons and Victoria some years earlier than that on the Gold Coast, and it is even claimed by some that the phenomenally successful industry of the British colony was commenced with a seed pod obtained from Ambas Bay

There is probably no single feature in colonial enterprise which can compare with the cocoa romance of the British colony of the Gold Coast. The honour of having introduced the industry into that colony is eagerly debated. Everyone agrees that it belongs to either the Basel Mission through their introduction of West Indian Christians, or to a certain native carpenter returning from Ambas Bay, or Victoria. Mr. Tudhope, the Director of Agriculture, is inclined to give credit to the native, but it must be admitted that the Basel Mission authorities possess the most circumstantial evidence in support of their claim. One of their oldest missionaries at Christiansborg states that about the year 1885 he saw the original cocoa tree at Odumase ; another, that he saw this tree in full bearing in 1895. It is instructive to recall that the first export, amounting to 80 lbs. weight, was in the year 1891—that is six years after the original tree was seen at Odumase.

The missionaries, however, readily admit that soon after their agents introduced cocoa at Odumase, a native arrived from the Cameroon colony and planted beans at Mampong. From these two centres, fifteen miles apart, the industry has established itself in every district of the colony and penetrated ten days' march beyond Kumasi.

The organization is of the simplest kind—purely and solely a native industry, few of the plantations being large ones, none more than about twenty-five to thirty acres and the majority not more than two to five acres. We saw none owned by white men, although I believe there are one or two, which are, however, quite insignificant. The volume of cocoa which pours out from the Gold Coast colony flows almost exclusively from countless small holdings spread all over the hinterland. The farms are not so close together as those of San Thomé, but the traveller cannot walk many miles anywhere without passing through the plantations of cocoa and palm trees.

The atmospheric conditions resemble the Mayumbe country and San Thomé, the rainfall varying between

32·09 and 54·92 per annum, otherwise the territory is not so well watered as the Belgian and Portuguese possessions. In spite of this, the colony can produce a quantity and quality of cocoa that compares well with other areas. When at the Botanical Gardens of Aburi, we saw a plot of cocoa measuring one and two-fifths acres with 259 trees planted fifteen feet apart. The yield from this plot between October 23rd and December 31st, 1909, was 18,200 pods. Mr. Anderson, reporting upon this experimental plantation says, " Such results will not often be exceeded in any cocoa-growing country."

In the year 1891, we almost see that Gold Coast native offering for sale the first harvest of cocoa. It is only 80 lbs. in weight and with the greatest ease he carries it to the white man's store. To the amazement of his native friends the grower received £4 for that basket of cocoa !

Twenty years later the export of 80 lbs. weight has grown to nearly 90 millions. Since the day that the native husbandman disposed of his 80 lbs. of cocoa, the industry has never wavered. We were informed by white men who have been long on the coast that when the natives realized the value of cocoa there was an impetuous and overwhelming demand for seed until competition became so keen that a sovereign a bean was the general rate !

In 1902 the export had exceeded £100,000 ; in 1907 it had passed half-a-million, and in 1911 leaving gold in the rear of competition for first place it raced away beyond the finger post of a million and a half sterling. The whole of this, be it remembered, is a native industry !

The Gold Coast natives are justly proud of their

extensive enterprise and assert that they will not cease extending their plantations until every acre they can cultivate and every man they can use is producing cocoa.

Not the least interesting spectacle in the Gold Coast is the transport of cocoa, the bulk of the inland produce being carried by porters to the railhead, and sometimes the roadways as far as the eye can penetrate are one long line of cocoa bags on the heads of hundreds of carriers. This carrying trade has produced an extraordinary flow of free labour into the whole hinterland of the Gold Coast. At Adawso, a buying station nearly fifteen miles from the railhead, one firm alone employs in the season over 3000 carriers who cover the distance to the rail station of Pakro once, frequently twice, a day with a bag of cocoa. The remuneration being according to the quantity carried, there is an eagerness to earn the maximum within the twelve hours of daylight. The men who leave by daybreak will return about three o'clock in the afternoon, often to pick up another load and carry it to the railhead, returning again by moonlight.

The carriers are mostly Hausas, but the fame of the Gold Coast carrier traffic has spread far into the northern regions of Africa with the result that recognized caravan routes now come right down through the northern territories. These carriers, many of them from around and even beyond Lake Chad, drive herds of cattle down to the Gold Coast colony about harvest time. They sell the cattle and then carry cocoa for the season. When the main harvest is over and there is little cocoa carrying, they will purchase loads of kola nuts which they carry back with them to the far interior and sell *en route* at a considerable profit. Thus they make a threefold financial return—on the sale of cattle, cocoa carrying, and profits on the kola nut trade.

The transport of cocoa is chiefly in the hands of alien labour, and should the flow of this labour cease from any cause whatever, the cocoa industry would suffer a check from which it would take years to recover. The coastal regions are fairly secure, for most of the districts within twenty miles of the coast are reached by a daily service of motor lorries under the management of the European cocoa-buying firms. Many of the native farmers within thirty miles of Accra, however, with true African trading instinct prefer selling their cocoa at a higher price at the port of embarkation, and so have created the interesting system of " barrel rolling." In the season these strongly bound and ponderous casks are purchased from the European stores, filled with cocoa, and rolled to the sea-shore. Travelling along the somewhat primitive Gold Coast roads one meets at frequent intervals perspiring natives struggling with the barrels which, filled with cocoa, weigh considerably over half-a-ton. They may be " holding on " to a barrel racing down a steep incline, or three of them straining their utmost to force the ponderous weight up a steep hill. Occasionally they come to grief, for we saw more than one cask which had fallen over a cliff into a deep gorge below. Generally speaking, three men will undertake to roll two barrels to the coast, the three concentrating their efforts upon a single barrel going uphill, while on the level road or down hill they control the two barrels between them. We met three such men who had rolled

COCOA DRYING IN SUN.

two casks for twenty-five to thirty miles, a task of two days, for which they receive 20s. per cask.

The problem which faces administrator, merchant and native producer is that of transport. This threatens to become acute, for we were informed by a merchant who recently journeyed beyond Kumasi that large consignments of cocoa were lost owing to the lack of transport facilities. At the same time, given a fair price for cocoa in the home market, just treatment for transport labourers, the extension of roads and light railways, there is no reason why a single ton of cocoa should fail to reach the coast.

In the Gold Coast colony the white man occupies his normal position in the tropics—the connecting link or middle-man between the European manufacturer and the native producer. The Government very wisely endeavours to keep the industry in the hands of the native farmers and assists them by sending lecturers through the colony, whose duty it is to advise the farmers upon pruning, fermentation, drying, the danger of pests, and the general principles of modern agricultural science. With inherent instinct, the British Government recognizes that the real asset of the colony is the indigenous inhabitant, whose material and moral progress is not only the first, but the truest interest of the State.

The other British colony in which cocoa has a future is Southern Nigeria. To read the Government reports of ten years ago there seemed little hope that the natives of this colony would become cocoa farmers, or indeed that they would ever do much more than vegetate in the agricultural world. Africa is the land of surprises, and

more and more the African is surprising Europe by
exploding the lazy nigger theory."

The Acting Secretary of Southern Nigeria, writing
his 1903 report from Old Calabar, said :—

"With every year that passes, it becomes in-
"creasingly important that new exports, indicating
"new areas of work and development, should make
"an appearance on the export lists of the Protec-
"torate. That 'Palm Oil' and 'Palm Kernels'
"will ever cease to be the dominant products is more
"than unlikely ; but these products demand nothing
"from the native in the way of labour that the
"veriest bushman cannot carry out. Portions of
"this Protectorate must be gradually turned over—
"and education may succeed, where persuasion
"fails—to the production of other commodities. It
"is not in the nature of the average West African
"to lay out capital for which there is no immediate
"return. He can understand the yam growing at
"his door ; he can understand the cask of oil to be
"filled before his ' boys ' can return with the required
"cloth, pipe, or frock-coat, but he will not sow for
"his son to reap ; nor will a village work, of its own
"initiative, for the benefit of the next generation
"that is to occupy it. It is this difficulty that has
"rendered so great the task of encouraging the
"rubber industry. It is for this reason that cocoa
"and coffee have never been properly taken up by
"the natives themselves."

This is just what the Belgian and German Governments are proclaiming to-day.

At this period cocoa was just beginning to grip the native mind in Southern Nigeria ; he had begun to " sow for his son to reap " ; he had begun to understand something more " than the yam growing at his door " ; he had in fact just dispatched 300,000 lbs. of cocoa to Europe. The very next year the Acting Governor was able to write : " There has been an enormous development in cocoa," and the Southern Nigeria natives, as if in unconscious protest against the Governor's 1903 report, poured into the European markets over 1,000,000 lbs. of cocoa beans ! Two years later, the export had risen to 1,500,000 lbs. Turning to the Government report three years later again, we find that the export had again doubled itself, and was then over 3,000,000 lbs.   " These figures," said the Colonial Secretary, " indicate the extraordinary expansion that has taken place of late years in the cultivation of this plant." Finally, turning to the most recent report, we find that the export has again doubled itself in two years, *i.e.* over 6,000,000 lbs.

The actual figures are as follows :—

1903	.	.	.	.	288,614 lbs.	£3,652
1904	.	.	.	.	1,189,460 ,,	£18,874
1906	.	.	.	.	1,619,987 ,,	£27,054
1908	.	.	.	.	3,060,609 ,,	£50,587
1910	.	.	.	.	6,567,181 ,,	£100,000 (approximately)

It is somewhat doubtful whether this ratio of doubling the output every two years will be sustained, for it is considerably in excess even of the Gold Coast rates of increase. There are advantages possessed by Southern Nigeria which natural conditions deny to the Gold Coast—the heavy surf, and the lack of good shipping accommodation, tell heavily against the merchants and the native producers of the Gold Coast, whereas it is possible to load and unload cargoes in Lagos without their suffering any damage from sea water. Again, the cocoa areas of Southern Nigeria enjoy in the main a more generous water supply than those of the Gold Coast.

The general statistics of the cocoa trade, compiled upon the materialistic basis of tons and sovereigns, are not without interest to the man outside the cocoa community. For example, the Portuguese at present produce more cocoa on their two little islands of San Thomé and Principe than any other cocoa-producing area in the world. They produce from those 400 square miles of volcanic rocky land more than twice the quantity produced by the Republic of Venezuela with a tropical region of nearly 400,000 square miles. At the same time out of the eighteen cocoa-consuming countries of the world the Portuguese are proportionately the smallest consumers of Linnæus' "Food of the Gods." Another interesting feature is the growth of the British export from the West African colonies. Within ten years this has multiplied itself something like twelve times over, i.e. in round figures from about 2500 tons in 1902 to over 30,000 tons to-day.

Cocoa grows apparently with greater ease in West Africa than in any other cocoa-producing area in the world. The elaborate systems of manuring which seem imperative in most tropical colonies never enter the head of the West African producer. He piles the fermenting husks in heaps between the rows of trees and then when thoroughly decayed he throws the refuse round the base of the trees.

Insect pests abound, in fact it is seldom one sees a cocoa tree free from the tunnels of the devouring termite, and the bark-boring beetle too makes his presence felt, particularly in the German Cameroons, but in the great cocoa-producing colonies of the Gold Coast and San Thomé, the natives and the Portuguese are profound believers in the principle of " live and let live," at least in favour of the insect world. The Germans, in all things scientific, have attempted to deal with the pest-ridden area by manuring with superphosphate and potassium chloride, and a largely increased yield is claimed for areas treated in this manner.

In very few plantations that we visited was there any adherence to the wide spacing so strongly advised by expert agriculturists. The British Botanical Gardens of Aburi set an example by laying out experimental plots with cocoa trees fifteen feet apart, but the natives in that colony, and also in Southern Nigeria, ridicule this advice and declare that at such distance they find the rays of the sun are able to penetrate so freely that the ground becomes baked and the roots are robbed of the humidity which is vital to the growth of good cocoa trees. It is noteworthy that on Grenada and other West Indian estates, there is also a tendency to plant more closely than the experts advise. Neither in the Gold Coast nor in Southern Nigeria do many plantations give wider spacing than eight feet apart, and thus many of them crowd from 500 to 700 trees upon a single acre. The plantations in British West Africa being entirely under native control, there are no very reliable statistics upon the annual yield per tree. One official at the Botanical Gardens of Aburi estimated that the natives obtain about 7 lbs. of cocoa per tree per annum ; this is a very high average, and I am inclined to think seldom attained, for in Trinidad the annual yield is somewhere about 1 lb. per tree. We visited one large cocoa plantation in Southern Nigeria, where a native had planted 100,000 cocoa trees about one-half of which were already yielding, and from the 50,000 he had obtained within the year 30 tons of cocoa, or an average per tree of a little over 1¼ lbs.

The most important question, that from which the planter is never free, is that of labour. The Germans put the labourers on contracts of twelve months with wages of 8s. to 10s. per month with food, but the conditions of these plantations are not likely to inspire any great enthusiasm amongst humanitarians or economists. The abject fear exhibited by the natives whenever the white man approaches is too eloquent to be mistaken, moreover the whip is carried by the planters as openly as a man in Europe carries a walking-stick. Whips and free contracts seldom go together. Under another section I have dealt with Portuguese labour which in the main is a system of slavery, although it carries with it a paper wage of about 10s. per month and rations.

The native farmers of Southern Nigeria and the Gold Coast employ a good deal of native labour and generally speaking find little difficulty in obtaining all they want. These native farmers, however, prepare their contracts somewhat differently from the European, generally they are for " twelve months of thirty working days," and the wages vary from 15s. to 20s. per month, whilst a foreman will get 30s. a month. The labourers are free to go at any time, but those who complete their contracts to the satisfaction of the employer are usually given a bonus.

Cocoa growing is probably the least arduous labour in the tropical world of agriculture, as it involves less exposure and at no stage can it be called dangerous as is the case with copra, palm oil and indigenous rubber. The proportion of labourers employed varies according to the colony and circumstances. In the island of Fernando Po, the planters endeavour to employ one man per acre, but the restricted supply of labour seldom permits so ideal a proportion. In the Portuguese island of San Thomé, one labourer is allowed for each hectare under cultivation, but it must require a good deal of " persuasion " to get a native to control an average of at least 2¼ acres of cocoa.

Concurrently with native labour is the question of white supervision which is necessarily costly. On one cocoa plantation of San Thomé, with a total expenditure of £23,000 no less than £3000 is spent upon white control. Upon those Belgian plantations of Mayumbe which are cultivated by free labour, there is barely any white supervision, whilst on the Portuguese islands the proportion of employés works out at about one white man to every thirty natives.

So far as it is possible at the moment to forecast the future of cocoa production in West Africa, the British system alone rests upon a solid basis, for the obvious reason that all other fields are dependent upon systems of labour supply which have little chance of continuance, much less extension. The indigenous industry of the British colonies working in its own interests, unencumbered by the heavy cost of European supervision and the drawbacks of imported contract labour, will, under the guidance of a paternal and sympathetic administration, certainly outdistance and leave far behind in the race for supremacy such systems as those which prevail in San Thomé and Principe.

This virile British enterprise which is bounding forward throughout the Gold Coast and Southern Nigeria has only one real enemy—the concessionnaire hunter. Fortunately, the British Government is fully alive to the danger and is determined, so far as possible, to keep the agricultural land in the hands of the natives. If this can be secured without placing powers in the hands of the Government which would lead to widespread disaffection and unrest amongst the natives, then the cocoa industry of British West Africa promises to eclipse all other cocoa-producing areas of the world.

IV

THE PROGRESS OF CHRISTIAN MISSIONS

THE day has gone by when the world could dismiss
Christian missions in West Africa with a contemptuous
sneer, for Christian missionary effort with its eloquent
facts, definitely established, can no longer be ignored.
Of all the forces which have made for real progress in
West Africa, Christianity stands some say first, others
second, but none can place it last. To it belongs primarily
in point of time at least, the economic prosperity of the
Gold Coast. To it belongs, almost entirely, the credit
for the native clerks and educated men on the coast.
To it the natives owe their knowledge of useful crafts.
To one section of the Christian Church at least belongs
the honour of having on the spot saved the Congo natives
from extirpation.

Whilst all missions have much in common, the investigator cannot but observe the fact that administrators and commercial men alike will, in the majority of
cases, hold in a measure of contempt the Protestant
missionary, whilst they esteem highly his Catholic
brethren. One searches for a reason for this attitude,
which can neither be found in the devotion of the
missionary—for heroes abound in both sections—nor is
it to be found in the character and success of their

respective missionary labours, for in this particular both sections are witnessing encouraging results. The only answer which the administrator and trader will give is that Father O'Donnell is " a good fellow." It is difficult to escape from the conclusion that the good father is more " diplomatic " than his bluff and somewhat puritanical Protestant *confrère*. The Protestant missionaries with greater freedom than that allowed to the Catholic Fathers, criticize administrations, report abuses, and generally give any form of oppression or iniquity a quick, even reckless exposure. The colossal crime of the Congo was exposed on the spot almost entirely by the Protestant missionaries, although far outnumbered by the Catholics. In the French Congo are established several Roman Catholic Orders, yet hardly a priest has raised his voice against the atrocities committed there. The slavery of Angola and San Thomé has been exposed primarily by Protestants, the priests standing by and for the most part content to witness the traffic in human beings without a protest. I do not condemn, but merely state facts. I know too well how the sufferings of native tribes have appealed to generous members of the Roman Catholic Church, but no review of Christian missions in West Africa would be honest or complete without some reference to this fundamental difference between the two great sections of the Christian Church.

My chief reason, however, for calling attention to this feature is that the antipathy towards Christian missionaries is hardly likely to become less marked in the near future. The great changes which are taking place may precipitate a grave situation within the next twenty years. The attitude of administrators is no longer the benevolent tutelage of native races. There is an increasing autocracy in most colonies ; the martial spirit with its harsh regulations and rigorous discipline, so out of place in nature's calm paradise, is permeating every department of affairs. This spirit brooks no opposition, knows no sympathy, and sometimes even forgets justice. It blows hot or cold, where and when it listeth, but it tends always towards menacing native peace and progress. High-minded Christian men must be driven by this restless spirit into an increasingly resolute defence of their native communities.

Commercial methods, too, are undergoing a still more far-reaching change. As I have already pointed out, the old-time merchant is giving place to the highly organized syndicate, which possesses neither heart nor conscience and is generally strong enough in influence at home and power abroad to menace any administration, and, if necessary, threaten the various Governments in two, three and even more countries at one time. The missionary, bold in his isolation, knowing no higher earthly authority than his highly tempered conscience, willing, if need be, to suffer any extremity, is bound to find himself more and more in conflict with the exploiting energy of these vigorous dividend seekers. This conflict is of course an excellent tonic for the Church, but it makes the lot of these isolated men and women in Central Africa very much harder to bear.

The forces of Christianity have not yet made much headway in the far hinterland of the Sierra Leone Protectorate, the northern territories of the Gold Coast, nor in Northern Nigeria. In the Sierra Leone colony, where slaves liberated during a period of fifty years were dumped down as they were released by British battleships, Christianity has permeated fairly completely the life and habits of the people ; nearly two-thirds of the population are nominally Christian, whilst the Mohammedans number less than one-tenth. In the Gold Coast the traveller may witness some of the most effective missionary work in West Africa. The Basel Mission alone has over 30,000 adherents who find about £5000 a year towards mission expenses. Another notable fact is that the natives have invested in the Mission Savings Bank over £23,000, a sum considerably in excess of the amount deposited with the Government. As was the attitude towards the Quaker bankers of Puritan England, the Christian community of the Gold Coast is regarded by the natives as the safest repository for the wealth of both worlds.

In Southern Nigeria Christian missionaries find themselves confronted with a firmly entrenched Mohammedan community. Something over fifty per cent. of the population is Mohammedan, and that of a most attractive order. None can meet the leading Mohammedans of that colony without being impressed with their simple piety and their tenacity to what they regard as their invincible faith. Officialdom opposes the advance of the emissaries of Christianity in the more northerly territory, on the ground of trouble with the Moslem community. This attitude is regarded by most Mohammedans as anything but a compliment to their religious faith, holding firmly as they do that the Koran is powerful enough to withstand all the assaults of another creed. Below Nigeria, that is south-east of the Niger delta, Mohammedan influence is left behind, and Christianity is confronted with simple paganism. Not the blood-thirsty and strongly entrenched barbaric paganism which confronted Livingstone in East Africa, Ramseyer at Kumasi, Hannington in Uganda, and Grenfell in the Congo, but a paganism so broken by the forces of civilization, so rent and riven by internal mistrust, that the masses of the people are crying out : " Who will show us any good ? "

Efforts to win West Central Africa to Christianity divide themselves into two periods. The first effective efforts were made by the Portuguese and Dutch settlers in the sixteenth and seventeenth centuries. The first period was almost exclusively due to Roman Catholic zeal, which, under the blessing of the Pope, regarded the tropics as a preserve of the Vatican. The nineteenth century witnessed but little advance, until Livingstone's enthusiasm and his romantic career lighted a flame which spread throughout the civilized world, and Pro-testantism, awaking to its opportunity, began to pour missionaries into the tropical regions of West Africa. The Basel Mission attempted the Gold Coast, and its first missionaries perished to a man ; the Church Mis-sionary Society pushed on its work from Sierra Leone away up the Niger, where men and women did little more for a time than replace the dead and dying ; the Metho-dists, never behind any other denomination in enthusiasm, began work in Sierra Leone, Calabar, and the islands of the Gulf of Guinea ; the Baptists established an excellent

mission in the Cameroons, where they were " elbowed
out " by the Germans, and at a later date commenced
their great work in the Congo.

Little remains in the social life of Africa as a result
of the work of the early Roman Catholic missionaries.
The tribes have no settled church organization based
upon the devoted efforts of three centuries ago. Ruins
may be seen in several parts and extremely interesting
ones too. On the islands of San Thomé and Principe,
we frequently saw the partial structure of churches, one
of which must have been erected very early in the sixteenth century, for a tomb close to the chancel, grey
with age and moss grown, was dated 1542. If the
colonists of San Thomé were zealous slavers, they certainly gave much of their ill-gotten gain to the erection
of churches. Fragments of these edifices are lying about
in the tropical undergrowth and an examination will
show that marble pillars, façades, altars, even common
stone, had been gathered from the four corners of the
earth to build ornate " Houses of God " on these isolated
rocks in the Gulf of Guinea. Visiting one of these ruins
we were struck by the pathetic reverence with which
the natives regarded those crumbling walls ; the priest
had long since died, and there was none to lead those
almost hopeless souls along the path of religious faith.
Standing inside those four walls, gazing at the broken
altar and the creeper-clad walls, we were forced to keep
our heads covered, for the ruin had lost its roof generations before, and the equatorial sun was pouring its
direct rays upon us. Directing a question to some of
the natives standing near by, we were amazed to find

that they refused to answer; two or three times we repeated our questions, but they all maintained immovable positions and refused to utter a single word. A man close at my elbow then informed me that no native could reply whilst the white man kept his hat on his head in the House of God! The silent rebuke of those simple natives forced us to leave the precincts of the old ruin and pass into the little chapel which still remains more or less watertight. Into this place, not more than ten feet square, the natives had moved the images of the Virgin and Apostles, and in the centre of the room a native palm oil lamp sent forth its unpleasant odour. This lamp was half African fetish and half salvation to those natives, for their worship had degenerated into a sort of corrupt Zoroastrianism, and the Alpha and Omega of their religion seemed to be the uninterrupted burning of this light. They were most insistent that since the foundation of the church, between 1500 and 1530, the light had never been allowed to go out!

This, however, was but one testimony to the relic worship of the slave islands. Along the roadsides, in secluded corners of out of the way roças, nestling in plantain groves, the traveller may see miniature chapels constructed from rustic forest tree branches, very similar to the fetish houses of the mainland of Africa. In most of these one also sees little prayer-stools, and in all of them a rude cross roughly cut out with the native axe and the cross pieces bound together with forest vines. Most of these crosses are surrounded by native pagan charms, and thus all that is least essential in Christianity is joined together in native religious fervour with the superstitions of paganism, and this gives a melancholy impression of the result of the years of toil and sacrifice by men and women devoted to the theory of the Christian Faith.

Ichabod is written along every roadside and in every ruined chapel; the very images in decay seem to utter the word, and the mind is compelled to recall the fact that Christianity in creed only, without Christian practice, is foredoomed.   Surely the curse of the miserable slaves of generations ago rests upon everything on those islands; by their agony and bloody sweat they toiled to erect those magnificent churches, the crack of the whip on the slave plantations extorted the gold which purchased the images of the Virgin, to add lustre to countless churches and to purchase images of the compassionate Christ for the cross roads and public places.   One wonders what all this parade meant to the slaves at the time. They have long ceased to suffer the bonds of slavery, or the crack of the whip; those slaves whose toil built the churches and bought the crucifixes have gone, and though decay everywhere marks the one-time existence of an unholy Christianity, one element remains and flourishes —a slavery, without any hope beyond that which may be inspired by the hybrid of effete Christianity wedded to African superstition.

The results accruing to the second period of Christian propaganda have the unmistakable signs of a vitality which will revolutionize Central Africa.   Whilst purely missionary zeal centres itself upon the heroic figure of Livingstone, recognition must be given to Henry Stanley, and also—though one hesitates to couple the

THE CRUCIFIX IN AFRICAN FETISH HUT ON THE ISLAND OF SAN THOMÉ.

RUIN OF ONCE IMPOSING CHURCH ON THE ISLAND OF PRINCIPE.

name with these two heroes—to King Leopold. Looking back upon African history, one fact emerges above all others, that the work of Livingstone and Stanley together had created an international interest in the position of the peoples and the possibilities of the countries in those regions. This condition observed by King Leopold, his master mind promptly seized and exploited it. The crafty Belgian monarch saw that by preaching Christianity and civilization for the African, his long-awaited opportunity for colonial expansion and a place in history would be gratified,—a place in history he has that none assuredly will envy ; his people, too, possess a colony, and though they do not see it to-day, they will yet heap their curses upon the sovereign who has fastened the millstone round their necks.

The labours of Livingstone, Stanley, and King Leopold, culminated in the Conference of Berlin, which was unique in that it had for its programme not only the interests of honest commercial expansion, the suppression of the slave-trade, the sale of arms, ammunition and alcohol, but also that of stimulating Christian missionary propaganda, and by its subsequent treaty, missionaries were encouraged to win pagan tribes from barbarism. The immensity of the area which by this historic event was thrown open under international stimulus to the forces of Christianity is not generally realized. The Congo basin extends far beyond the boundaries of the Belgian colony. Its northern frontier reaches the tributaries of the Niger and the Nile, while its eastern border includes a large section of German East Africa, and in the south and west larger areas still of both British Central Africa and Portuguese Angola come under the operations of the Act of Berlin. In and around this great pagan area, almost as large as the European continent, the forces of Christianity have within the last half century been concentrating their energies.

Christian effort in these regions is confined to no single country, and is the monopoly of no single denomination. Great Britain, America, Germany, Sweden, and France have all found devoted men and women, and have all poured forth most generously the necessary funds. Anglicans, Roman Catholics, Free Churchmen, and Lutherans, have all taken their share, selecting spheres which for various reasons they considered themselves best able to manage.

The character of the work, however, differs considerably. At first Protestant missions revolted against the idea of industrial missions ; they had, and it must be admitted they still have, a constitutional objection against anything which provides a " return." It is difficult to find a reason for this, but probably it is due to a revulsion from the practices of Pizarro and his miscreants in Peru, and of the slave-dealing work of the Portuguese, in which the Church of Rome became so deeply involved. This dislike for any other work than that of simple preaching and teaching left to the Roman Catholics the whole field of industrial enterprise and right splendidly they have occupied it. There are many separate features which one dislikes, but looked upon as a complete work the Roman Catholic missionaries are rendering noble service to stable progress. I shall not readily forget visits to their farms on the Congo ; to their admirable outfitting, printing, house-building, and wheelwright departments of German Togoland. In Lome we saw a score of lads learning bootmaking under the patient tuition of a lay brother. In the tailoring shop another score were cutting out and making suits of every description, from the cheap 20-mark ducks to the 150-mark dress suit to which the superintending Father was putting finishing touches—and made for a native too !

If in earlier years Protestant missions hesitated to engage in remunerative industrial pursuits, they scored heavily over their Catholic *confrères*, and continue to score, in medical work. It was at first difficult to make the native see the advisability of even comparative cleanliness, for ablutions of any kind are, with many natives, a degrading practice only fitted for the effeminate white race. " What ! I wash ? " exclaimed an old chief to us in horror-stricken tones, when once I asked him to take a journey to the river before sitting near our table. However, as he proceeded to do a worse thing—scrape himself—I withdrew and apologized for the insulting suggestion ! There is some hope that the medical fraternity will in time bring the natives to realize the value of the bountiful streams which God has given them, though they may retort that the devil has filled them with crocodiles.

It is, however, certain that the tribes of Africa are beginning to value the generous and devoted medical work of the Protestant missionaries. Journeying up the Congo one day we had on board a chieftain who three months before had left his village for an operation at a

mission station hundreds of miles below his home. The
senior missionary in this man's district had persuaded
him to take the journey and run the risk. The man had
been bedridden for years with an elephantiasis growth ;
his wives had forsaken him and most of his friends had
abandoned him. He had long given an obstinate refusal
to the missionary's proposal, but ultimately he was prevailed upon to make the journey to the distant mission
post. The day for departure came, and with it funeral-
loving friends, and weeping women who made the track
echo with a monotonous death wail as the man was
carried on board the steamer,—never, as they believed,
to return alive. Two months later the man had come
through the operation and seemed to be in perfect health.
He boarded the steamer in full vigour, carrying his own
box and sundry goods which the travelling native collects
from the long-lost brothers and cousins whom they have
a habit of discovering in every town. After three weeks'
steam, we were nearing the chieftain's home ; what a
dressing of the hair and anointing of the body took place
during several hours before the village itself was sighted !
Within hail, lusty voices shouted to the villagers that
their chief was aboard and was well and strong. The
cry passed from lip to lip until the beach was lined with
incredulous natives, the most hopeful amongst them
anticipating nothing better than that the man would be
carried ashore. Fifteen minutes later the ship was at
anchor, the " gangway " run ashore and lo ! the first
man to stride off the ship was the erstwhile bedridden
chief ! It was too much for the majority who promptly
took to their heels and bolted to a safe distance ! In a

few minutes, however, they realized that it was not a spirit, but the real man returned alive and well. Gradually they surrounded him, questioned him, gesticulated excitedly, rang the drums to inform the countryside that so great a miracle had taken place, and generally made such a din and noise that it was only with difficulty conversation became at all possible. That sort of sermon is far more eloquent to the native than many discourses on Christian ethics preached with the inevitable limitations of a foreign tongue and at the best often misunderstood ; moreover, it renders him very receptive to Christian teaching.

The advantage of medical work in Protestant missionary propaganda has indeed been great. But it does not stand alone, for the natives have of recent years witnessed and wondered at another spectacle—to them no less miraculous—white man opposing white man on their behalf. It is a grave misfortune to Christianity, and to the Roman Catholic missionaries themselves, that they have hitherto been unable to make common cause with their Protestant brethren in protecting natives from oppression. There is, however, some hope that this feature is passing away and that the future will witness their co-operation with those who fight and struggle for native freedom, for at present the *prestige* which accrues to the championship of native rights belongs almost exclusively to the Protestant communities. How powerfully this has operated was brought out in the report of the Commissioners, whom King Leopold was compelled to send to the Congo, in 1904. Writing in this connection, Monsieur Janssens and his Committee said :—

" Often, also, in the regions where evangelical
" stations are established, the native, instead of
" going to the magistrate, his natural protector,
" adopts the habit, when he thinks he has a
" grievance against an agent or an Executive officer,
" to confide in the missionary. The latter listens
" to him, helps him according to his means, and
" makes himself the echo of all the complaints of a
" region. Hence the astounding influence which the
" missionaries possess in some parts of the territory.
" It exercises itself not only among the natives
" within the purview of their religious propaganda,
" but over all the villages whose troubles they have
" listened to. The missionary becomes, for the
" native of the region, the only representative of
" equity and justice ; he adds to the ascendency
" acquired from his religious zeal the *prestige* which,
" in the interest of the State itself, should be in-
" vested in the magistrates."

Without doubt the advent of the late King Leopold
as an Administrator in Central African affairs was a
calamity almost impossible to exaggerate and had his
influence continued it would sooner or later have overrun
the surrounding territories administered respectively by
Britain, France, and Germany. That they indeed suffered
contamination was only too clearly demonstrated in the
case of French Congo, while German Cameroons was not
altogether free from the Leopoldian taint. On the Congo
itself, the very name of white man was made to stink in
the nostrils of the native tribes for all time, by reason of

the enormities in which King Leopold figured as the chief
actor.   But even that wily monarch outwitted himself ;
by his protestations of Christianity and Philanthropy he
was bound by the clauses of the Berlin and Brussels Acts
to countenance and encourage missionary enterprise, and
in practice to admit to the vast regions of the Congo
Valley the Heralds of the Cross.   And this was his un-
doing, for thereby came those exposures of almost incredible abuses, which shocked the civilized world, and
branded the arch culprit for all time as a murderer of
millions.   The same fatal blunder in his diplomacy
worked on the spot salvation for the remnant of the
people.   They flocked from all quarters to the protection of the missionary, who was to them the personi-
fication of justice.

What wonder that the word " Ingleza " (English)
became a passport to any native community, no matter
how wild and how averse to the white man.   It is recorded that the Belgian rubber merchants, recognizing
this, have sought safety when travelling amongst hostile
tribes in adopting the name and manner of the Englishman.   A certain Belgian tells how two of his colleagues
when travelling were attacked by infuriated natives
whose relatives had suffered at the hands of the rubber-
mongers, and on being told that it was the natives' intention to first mutilate them, as they themselves had
been mutilated, and then to put them to death, one of
them in his extremity sought refuge in the reputation
of the missionary and replied, " What, put Ingleza to
death ! "   While stoutly repudiating the assertion that
they were English, the natives requested them to sing

a hymn, and, fortunately for the desperate men, one of them remembered and sang a verse of a hymn he had learnt somewhere, and so amazed the natives that they let them go unharmed.

" Ingleza nta fombaka " (the Englishman never lies), has passed into a proverb and is spreading not only throughout the Congo, but even into Portuguese Angola. Possessing the unbounded confidence of the native mind, the Christian missionary, reinforced by practical medical work, may, if he desires, possess the vast unoccupied fields of the continent and obtain there an ever firmer foothold.

Within recent years, however, Protestant missions have taken up with increasing zeal industrial and commercial enterprises in the interests of the natives. We were unfortunate in being unable to visit what I am told is one of the finest industrial enterprises in West Africa—the Scotch Calabar Mission, but apart from those of the Roman Catholics we inspected several Protestant establishments. The British Government, recognizing what is now becoming common ground, that a purely literary and spiritual education does not produce the most robust type of civilized African, is now combining technical training in industries with literary studies, and no longer gives grants of lump sums to missions, but so much per head for the " finished product," e.g. a native attaining a given literary and technical standard. In the Gold Coast the maximum per annum is 27s. 6d. per capita. In a school at Christiansborg, the annual upkeep of which costs £500, over £170 was earned in one year by the ability of the scholars in this way.

The Primitive Methodists have a very effective little Industrial Mission on the Spanish island of Fernando Po. Under the vigorous and enlightened leadership of the Rev. Jabez Bell the mission situated at Bottler Point is now so prosperous that the returns from the cocoa farms together with subscriptions from the native members, more than cover the expenditure. If in any forthcoming rearrangement of the Map of Africa Fernando Po should come under Germany the character of the Primitive Methodist Mission on that island is bound to appeal to the practical-minded Teuton.

The price which Christian missions have paid for religious work amongst the pagan tribes of West Central Africa can never be correctly estimated. In the Congo alone Protestant missions have spent nearly one and a quarter millions sterling within the last twenty-five years. Out of some 550 missionaries, over 170 have gone to an early grave, many not living six months, some only a few days. These men and women were not only the matured youth of their countries, but they were compelled to pass the most rigid medical examination prior to acceptance by the missionary boards. They were indeed the flower of the Christian Church; moreover, the very difficulties and dangers which were known to exist, served to attract none but the strongest characters. Some people, incapable of recognizing sterling qualities in any but themselves, have written and spoken of missionaries as those who could not have made their way in any other sphere of life. Whatever may be true of other mission fields, so far as the missionaries of West Africa are concerned, the majority resigned good and assured positions and accepted a comparative pittance in order that they might serve what surely is the greatest of all causes. I have failed to obtain statistics from the Roman Catholic Church, but the foregoing applies equally to the devoted men of that body. With them, as with the Protestants, it has been *via crucis via lucis*.

The following statistics, so far as they are a guide to Christian progress, show some of the results achieved by the missionary forces of Protestantism in West Africa :—

	Adherents.	Scholars.	Annual Native Contributions.
Sierra Leone			
Anglican	12,700	3,283	£7,267
Methodists	7,584	2,665	—
Nigeria			
Anglican	40,700	15,089	£11,676
United Free Church . . .	6,431	3,675	£2,834
Methodists, including French Dahomey, German Togoland, and Fernando Po .	7,137	3,793	—
Gambia			
Methodists	1,058	594	—
Gold Coast			
Society for the Propagation of the Gospel	3,273	—	£677
Methodists	61,481	7,821	—
Basel Mission	35,000	—	£9,500
Congo			
Baptists	4,536	11,637	—
American Baptists . . .	5,230	7,500 (est.)	—
Presbyterian	10,000	8,000	
Swedish	1,821	5,721	—
French Protestants . . .	1,800	1,000	—
Angola			
Methodists	750	1,083	£325
Other Missions in West Africa Estimate	15,000	8,000	400
Totals	214,501	79,861	£32,679

From the statistical tables of the Protestant Missions, we have a known membership and communicant list of over 200,000 men and women, and nearly 80,000 scholars under daily Christian instruction. If to this be added an equal number in connection with the Roman Catholic Church—probably a generous estimate—West Central Africa possesses a Christian Church of something approaching half a million strong. This, however, does not take into account the large native interest in Christianity evidenced by the considerable purchase of the Scriptures. Every year the British and Foreign Bible Society ships some thousands of pounds worth of Bibles to the different colonies, the natives contributing an increasing sum to the Bible Society, which gives a " return " in cash from the native Christian community of the Protestant Churches of over £30,000 per annum, or an average contribution of over 4s. 3d. per head throughout the Churches.

The fact that the results of missionary industrial enterprise are hampered by a not unreasonable dislike to " profit-making " prevents embarkation upon those bye-products of industrial activity which render commercial enterprise financially sound. A missionary is usually quite willing to teach men to adze timber, plane boards, square joints, lay bricks, and grow cotton and rubber, but he knows that his Board and its supporters regard " profit " with a very critical eye. Richard Blaize, an educated native of Abeokuta, left his fortune to meet this difficulty and now extensive workshops are erected at Abeokuta, and all the public buildings of that splendid city have been erected

" at a profit " by the Christian Industrial School of Abeokuta.

In the Gold Coast the German Basel Mission leads the way with engaging vigour in the matter of industrial missions. The commercial section of the Mission includes industrial training institutes, and nothing could be more pleasing than the interest and energy with which the natives devote themselves to cabinet work, coach-building, and agricultural pursuits ; but the main activities of this department are those of the ordinary African merchant with the exception that the agents are forbidden to sell spirituous liquors. This branch of the work, which is conducted by twenty-three " mercantile " missionaries, is in every respect admirable. One of the leading railway managers remarked to me that, " The most business-like commercial house in the colony is the Basel Mission ; their men always know how many trucks they will require, their trolleys are to time, their goods properly bagged and labelled, and their whole organization so smart and up-to-date that they never dislocate the traffic." There can be little doubt that the attention given to business by the representatives of the Mission is due to the type of white men they can command—none are accepted unless they agree to make their employment a matter of conscience, and develop their commercial undertakings with the same motive as that which animates their spiritual brethren, with whom they share all things in common, with the exception of salaries, those of the mercantile brethren being considerably higher and based, to some extent, upon returns. The white agents are assisted by coloured

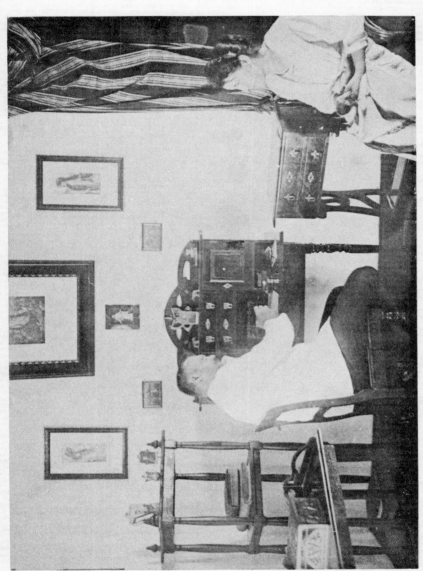

INTERIOR OF MISSIONARIES' HOUSE. BASEL INDUSTRIAL MISSION. FURNITURE MADE BY GOLD COAST MISSION SCHOLARS.

men in charge of branches, many of whom can show a record of service extending from 12 to 15 years, and some of them are now drawing salaries—including commission—of £500 per annum. These men are to be found on Sundays teaching in the Sunday schools, and preaching at the out-stations of the Mission.

The capital for these operations is derived, in the main, from three sources : (1) the Basel Mission itself ; (2) shareholders connected with the Mission, whose dividends are limited to 5 per cent. per annum ; (3) from funds in the Mission's Savings Bank, into which the natives of the colony have placed for security considerably over £20,000 at interest varying from 3½ to 5 per cent.

The results of the Mission's work can be seen all over the colony ; the polite native clerks, the managers of stores, the English-speaking planters, the coloured Government officials have nearly all of them received their training at the Basel Mission schools, and the Acting Governor does not hesitate to recognize that his best officials have been produced by the Mission. Testimony of this nature is unhappily seldom forthcoming from other colonies.

The industrial section usually executes orders to the value of about £4000 per annum ; its go-carts, trolleys, traps, and waggonettes are sent into almost every colony from Sierra Leone to German Cameroons. The net profits of this department average slightly over £400 per annum.

The commercial department is certainly one of the most profitable enterprises in the colony, and the stores of the Mission are crowded with purchasers throughout the day. The exigencies of business naturally precluded the possibility of obtaining with any degree of exactness the volume of trade done by the Mission, but some of the figures are eloquent testimony to the confidence the natives have in these mercantile missionaries. In the year 1909–1910, the Mission exported 35 tons of rubber, 14,000,000 lbs. of palm kernels, 600,000 gallons of palm oil, and nearly 17,000,000 lbs. weight of cocoa beans.

The profit-bearing transactions of the Basel Mission cannot be much under £150,000, which on the moderate basis of 8 per cent. net profit would provide the Mission Exchequer with a sum of £12,000 per annum. Government grants-in-aid of educational work amounted in 1910 to £240. There are also periodic collections in aid of Mission funds ; the native Church at Nsaba, for example, collected £240 last year. The whole expenditure of this Mission must be almost, if not completely, covered by its income from the various operations.

Whatever the actual financial position of this Mission, its general business operations, splendid educational institutions, its devoutly spiritual atmosphere, combine in forming one of the greatest—if not the greatest —force for progress in the Gold Coast colony. But the price has to be paid, for, according to the report of the Acting Governor, " The highest death-rate was again amongst the missionaries ! "

The future of Christianity in West Africa is hopeful but it has its dangers. First its very success may lead to disastrous consequences. In the early years the mission work was almost entirely in the hands of the extreme evangelical section of the Church, who subordinated everything to the actual work of preaching. We understand and sympathize with the fiery zeal that believes in doing all the preaching, but the native thinks the preacher a strange being, and frequently does not understand two sentences of Anglicized Bantu, or worse still, his Bantuized English! Circumstances have broadened the outlook and men are beginning to realize the value of training the native to do the preaching, contenting themselves with an apparently more restrictive sphere in the class-room and study. The native preacher thus prepared is zealous to a degree, and that he is ready to suffer incredible hardships and even torture, we know from the romantic history of the Uganda Mission. He is willing and able to carry his message further afield than the white man could ever hope to do ; he is, moreover, able to present his message through the medium of a complete mastery of the native tongue. The results of this form of propaganda are becoming almost startling. Christian evangelists from one territory are meeting those of far distant regions and in this manner the whole of the riverine systems of Central Africa are coming rapidly under the influence of Christianity. It is in this respect, rather than in tabulated statistics, that one sees the onward march of the Christian Faith. The bush native no longer clings to and prides himself in paganism ; if he is not a Mohammedan, he will tell you he is a Christian, even though his life and conduct would shut him out of the formal communion of any Christian Church.

This condition of affairs may lead to a grave situation, for already in several colonies the natives are restive under an inadequate white control or leadership. Educated in the principles of liberty, but without much respect for, or belief in, the nobler tenets of the Christian Faith, they are breaking away from Christian government and forming themselves into Christian communities in which personal desire is never allowed to conflict with accepted standards of ethics. One day I visited a leading " Christian " in a certain colony ; he showed me round the district, took me over his delightful little farm, pointed out his model dwellings, machinery houses, and so forth ; then I inspected a building with three compartments and was informed that one section was used as a " gin store," the middle section for prayer meetings, and in the third the man kept his wives ! All this he boldly asserted could be justified by reference to the Scriptures. I was not prepared to contest the assertion, because my host claimed his own conscience as the final arbiter of interpretation. The extent to which these secessions may go can be gathered from the fact that one such seceding church in West Africa claims a membership of over 10,000 adults.

The missionary societies, unable to supply sufficient men to cope with these vast areas, are forced to leave the movement almost alone and thus it spreads, and will continue to spread, until Central Africa is completely brought under the influence of a form of Christianity which for many years will be a caricature of the religion of Christ. The only hope, and happily a probable development, is that the religious wave, which is now moving irresistibly across the central regions, will be followed by an ethical wave which will give the " Light eternal " to the Dark Continent.

PART V

I.—The Map of Africa re-arranged.

I

THE MAP OF AFRICA RE-ARRANGED

FOR some months past eminent publicists in Europe have been busily engaged in rearranging the map of Africa in the interests first of one Power, then of another, but the unfortunate native has found scant place in these arguments. The only question which seems of any significance is the " price " this or that Power will pay for a given slice of the African continent. It would be rather interesting surely to know what the natives themselves think of the proposed change. Some of them have strong views upon the morality of disposing of other people's rights, to say nothing of treaty obligations which they obtained when agreeing to European sovereignty.

Four territories are, so to speak, in the melting pot of political speculation—British Gambia, French Congo, Belgian Congo, and the Portuguese colonies of East and West Africa. None of the Powers in control of these territories desire to add another foot of tropical Africa to the burden they carry already. Great Britain has a full share of responsibilities in the African continent. France, Belgium and Portugal, even if they desired to enlarge their tropical dependencies, have not yet established a case for expansion. Quite the reverse. One Power alone—Germany—is not only capable but apparently desirous of adding to her colonial possessions.

Though Great Britain may have no wish to add to her responsibilities, her people cannot allow her to relegate any portion of them to another Power. British Gambia must never be allowed to pass into the hands of France, and the quicker the French Government is told this, the better, not only for Anglo-French amity, but for the tranquillity and progress of our oldest African colony. Thanks to our merchant adventurers and the goodwill of the natives, Gambia was able to hold her own against the Portuguese, French and Dutch, until the British Government assumed direct control.

The British Government made a solemn agreement with the native tribes for a Protectorate over the Gambia colony, and " this agreement shall stand for ever." The British Minister who sets that agreement aside will be guilty of a crime against the fair fame of his country, but more especially against the loyal natives who so implicitly trust the unalterable promise of an Englishman.

The foregoing ethical argument cannot fail to find an echo in the breast of every Britisher, but there are Imperial reasons which reinforce that argument. The Gambia river possesses a draught of 30 feet right up the river to a position of twenty miles beyond Bathurst, providing altogether forty miles of deep waterway. I am told this is the only safe anchorage for a British fleet in West Africa, safe, that is, from attack. This little colony of 4000 square miles is contentedly following on in the path of progress, its inhabitants are loyally and affectionately attached to the British Crown. Only one thing troubles the population of Gambia and that is the periodic rumours of a transfer to another Power.

A categorical and clear statement from a Minister of the Crown that no such transference is contemplated or would be entertained is the least the native has a right to expect, and the Empire to demand.

Next in order of discussion has been the question of Portuguese colonies, and it has been mooted more than once that these should, either as a whole or in part, be transferred by Portugal to Germany for a financial consideration. The territories in question comprise Portuguese East and West Africa together with the islands of the Gulf of Guinea ; the area of the mainland possessions being 778,000 square miles and that of the islands 460 square miles, making a total area of 778,460 square miles.

No experienced Power would be prepared to purchase, even if Portugal would be prepared to sell, a portion only of the Portuguese possessions, because the several colonies properly managed dovetail into each others' requirements in such a manner that a separation of either would in all probability spell ruin to all. The richest of the colonies is that small island of San Thomé, but it cannot maintain its financial prosperity unless fed by labour from the mainland colonies of East and West Africa. Then the Angola finances are nearly balanced by the financial position of the cocoa islands.

Another argument is put forth, to the effect that the Portuguese treatment of natives demands a transfer of the territory to some more progressive Power, such as, for instance, Germany. Are those who advocate this policy quite sure that the " Progressive Power " would treat the natives better than the Portuguese ? if so, where is

the evidence ?   Does East Africa provide it ?   Does the
treatment of the Herreros and the shooting of British
Kaffirs demonstrate it ?   Those who support a transfer
on this ground should not lightly pass over these and
similar incidents.   They can be absolutely certain of
this one thing, that Germany would be " progressive "
enough to see to it that the cocoa plantations of San
Thomé had an abundant supply of labour—no matter
what the consequences to such a subordinate issue as
labour conditions.

Then I am told that " in any case the condition of
the natives as a whole could not be so bad under Germany
as under the Portuguese."   To this I cannot agree, for
though I admit that a number of colonial Portuguese are
slave traders, and that slavery is prevalent on the mainland and on the islands, I must, in fairness to the Portuguese, point out first that the Portuguese have a kindly
nature to which one can appeal, and secondly that signs
are not wanting of an awakening of the conscience of the
Portuguese nation in a manner which may lead to a
thorough cleansing of the colonial possessions of the
Republic.   Thirdly, there is no colour-bar in the Portuguese dominions.

To this argument I get the reply, " Public opinion in
Germany would insist on the abolition of existing slavery,"
but this is an argument which has no shred of foundation
in fact ; Germany is, in many respects, a progressive
Power, but she has no philanthropic soul for the well-
being of native races.   A single word from Germany
indicating a willingness to co-operate with Great Britain
during the Congo agitation would have saved thousands,

if not millions, of lives. That word was never spoken, the Congo tribes were left to perish, and German public opinion maintained a cynical attitude until the end. A merchant or two rendered yeoman service, but they were as voices crying in the wilderness.

Let Portugal retain her colonies, and resolutely begin to purify their administration and abolish slavery, but she must do it quickly if she is to retain the goodwill of those—and they are still many—who would deplore her disappearance from the map of Africa. She has said with an intensity demanding appreciation that she will not dispose of her ancient colonies, and this courageous reply evoked a warm response from all her colonists who to a man are intensely patriotic, but if Portugal should refuse to abolish slavery, she cannot expect that her most powerful Ally will be allowed to maintain an Alliance valued to-day by many of us, yet viewed with increasing uneasiness by a large section of the British Public. No one wishes to utter a word which can be construed as a threat, but every one knows that there are paths along which no British Foreign Minister can lead, much less force the nation.

Belgian Congo figures largely in every proposal for a rearrangement of the map of Africa. It is claimed that Belgium has annexed more territory than she can safely administer ; certain it is that in annexing the Congo she did not take over an ordinary colony. When Great Britain assumed responsibility for her African colonies their virgin wealth was practically untapped ; the people inhabiting the colonies, as a whole, welcomed the advent of her rule, and moreover Great Britain had in all her territories, with the exception of Egypt, a free hand. To a greater or less extent this applies to all African Powers, other than Belgium. When Belgian annexation took place, the Congo was in every respect a " squeezed orange " ; not only so, but the administration of that territory must remain subject to the paternal control of the European Powers. There is another feature which should not be overlooked, and which may yet cause difficulty. In the event of a general insurrection in French or British colonies, or in the event of invasion, these Powers can at once bring in an outside coloured army, which can, if necessary, be reinforced with white troops. Belgium can do neither. Let any material section of the native army revolt, which, by the way, is the ever present fear of its officers, and the Europeans must run for their lives. They would call in vain for troops from the Mother Country, for by the Belgian Constitution the army may not be ordered abroad, and for other reasons European forces could hardly be used in the Congo. With the unification of language amongst the native troops which is rapidly taking place, with the ever increasing spread of knowledge as to the use of arms, this peril has been gravely accentuated within recent years.

Apart from these general—and some of them, remote— difficulties, there are existing reasons for believing that the extensive Congo territories are too heavy a responsibility for Belgium. The country is over eighty times the size of Belgium, a proportion to the Mother Country not by any means without parallel, but circumstances differ so widely that they remove the question from all comparison with any other incident of colonial expansion.

The only countries which at all compare with the Congo are Uganda and Southern Nigeria, but these have not suffered as the Congo has done from thirty years of the maddest form of exploitation since Pizarro plundered Peru.

Uganda is just about 100,000 square miles and the Imperial grant-in-aid during the last ten years in direct administrative assistance (excluding railway credits) is no less than £1,075,000, or an average annual grant-in-aid of £107,500. This, it will be observed, is approximately an annual grant-in-aid of a sovereign per square mile.

The Belgian Congo cannot be managed upon less than Uganda, therefore it should require grants-in-aid from the Imperial Exchequer for the next twenty years, of not less than £1,000,000 sterling per annum. To this, however, must be added the exceptional demands from which no colony escapes. The question, therefore, is this : Are the Belgians ready to invest a sum of twenty to thirty millions sterling in the Congo during the next twenty to twenty-five years ?

The Congo is thirteen times the size of Southern Nigeria ; and though the density of the Congo population is considerably less, the tribes are much more widely scattered, and therefore require almost the same measure of white supervision, particularly in view of the direct taxation which everywhere prevails. There is the further consideration that many of the Belgian " officials " are to-day necessarily occupied with work which in other colonies is rightly left to the merchant. In view of these considerations a *pro rata* white personnel would seem to be essential. In Southern Nigeria Great

Britain has from 800 to 1000 white men of all grades
engaged in its administration. Taking the lowest figure
as a basis, the Belgium Congo, with its 900,000 square miles
of territory, coupled with the varied enterprises of the
administration, would require over 10,400 men, but the
colonial authorities find the greatest difficulty in maintaining an official personnel of about 2000 !

With a few notable exceptions, the type of official
on the Congo gives but little promise of any really effective
and enlightened administration. Can the Belgian
Government find the men which the colony requires ?
None can say actually what she may be able to do,
although none will deny that up to the present she has
failed to find either the number or the class of men vital
to successful colonial government.

The question the Belgians would do well to ask
themselves is—whether it would not be wiser for them to
administer a smaller colony properly than to continue an
attempt to govern a vast region like the Congo basin,
which, in the very nature of things, would be an enormous
task even for the most affluent European Power.

If Belgium could retain the Lower Congo, or a considerable portion of it, and transfer almost the whole of
the upper regions to another Power, the indemnity she
might expect to receive would, in such a case, permit of
the development of the Lower Congo in a manner which
would absorb all the activity that she could throw into
it for generations. There are great possibilities in the
Lower Congo, possibilities unprejudiced by the difficulties
which obtain in the Upper Congo.

If the Belgians would agree to the disposal of the

major part of the Upper Congo, the problem which would then confront statesmen would be that of finding another Power willing to assume so large a responsibility. France has the reversionary right, but could not be expected to add to her already too great African responsibilities. Clearly Great Britain could not accept the burden in view of the lead she took in the work of securing reforms. Portugal cannot effectively administer the territories she already possesses. We are thus driven to look to Germany.

It seems to be everywhere accepted that Germany would be willing to spend men and money on the administration of a large tropical colony, but again, is she prepared to accept the task of governing the Upper Congo?

A readiness to do so could hardly be other than welcome to the European Powers, always providing that Belgium were willing to share her present burden with Germany.

Such a transfer would not only unite the German colony of Cameroons with the Congo colony, but also with German East Africa, thus giving to Germany a great and uninterrupted trans-African colony larger than that possessed by any European Power in Africa. But the fatal objection to this re-arrangement is that there is no short and easy route to the sea which would be essential to a German control of the Congo. Even if the Matadi-Leopoldville railway territory were transferred, which would be extremely unlikely, the German Government would know that it was only a temporary route to the ocean.

Is it impossible for the Powers of Europe to take a truly large view of the situation and by making an immediate sacrifice, which in point of fact would be advantageous to each Power concerned, and thereby place the peace of the world upon a lasting basis ? Is it entirely out of the question to persuade France for an adequate *quid pro quo* to transfer the major part of French Congo also to Germany ?

France knows how she has been drawn into an almost impossible situation in that wretched colony, she knows how difficult it is for her to cleanse and administer it upon lines which are in accord with the obligations imposed upon her by the Berlin Conference. If France and Belgium together could be persuaded to transfer the whole, or the greater part, of French and Belgian Congo to Germany, a comparatively quick route to the ocean would be possible for the upper reaches of the Congo basin. If these two Powers could be brought to agree to such a transfer to Germany, they would individually be immeasurably the gainers, they would secure the peace of the world, and they would thereby add a lustre to their names which neither time nor eternity could tarnish.

Experience has shewn us that commerce need have no fear, for Germany welcomes and treats fairly the commercial houses of all nations. No favour is granted to a German firm which is prejudicial to the representatives of other Powers, whereas to-day the notorious fact is that merchants refuse to extend their commercial enterprises in French and Belgian Congo owing to the restrictions and irritating dues and charges which are imposed.

The remaining difficulty and the chief one is that of the treatment of the natives. Doubtless if it were possible to consult them, they would in both colonies vote for a transfer, not because they know anything of German rule, but because they would hope that a change would not involve them in a worse condition than they suffer at present. German administration of French Congo certainly could hardly be more oppressive than the French Government permits to-day. In Belgian Congo the natives would probably be treated as humanely and probably more justly than at present. Finally it is hoped that the German administration will, with the march of time, become less rigorous in theory and more humane in practice. On the whole, both from the commercial and native standpoint, the Congo basin stands to gain by a transfer to the German Empire.

This transfer, joining as it would the Cameroons with German East Africa, would provide Germany with a single African colony of something over two million square miles in extent ; occupied by a population of from twenty-five to thirty millions of people. A fertile colony larger than our Indian Empire, and approximately the same square mileage as the total possessions of Great Britain in the African continent.

The paramount question is, of course, what *quid pro quo* could Germany give in return for a re-arrangement of the African continent which would provide her with so vast a domain? Belgium would probably be willing to accept a cash indemnity with the retention of such portion of the Congo territory as she could safely and effectively administer. France is the only difficulty ; but if France chose to be generous enough to part with French Congo, might not such a spirit find an echo across the Rhine ? To put this into plain language, would not this vast colonial expansion thus placed in the hands of Germany be worth a rectification of the frontier of Alsace-Lorraine—or at least the gift of autonomy?

Such a solution of a great African problem would give a ray of light and hope to the darkest regions of the " Dark Continent," whilst in Europe it would settle for generations the peace of the world ; surely a fitting monument to erect to the memory of the martyred millions of the Congo !

INDEX

A

Abeokuta, 101, 283
Absinthe and natives, 99
Aburi, 254, 261
Accra, 256
African, artisan, the, 109 ; doctor, the, 110, 120 ; educated, the, 108
Anglo-Portuguese Alliance, 174, 201
Angola, 169
Anklets, brass, 34
" Anti-Slavery Society," Portuguese, 138
Ants, driver, 18

B

Baketi tribe, the, 24
Bakuba   ,,   the, 29, 40
Bakwala ,,   the, 39
Bakwiri ,,   the, 252
Baluba   ,,   the, 27
Bangalla ,,   the, 27
Bangalla River, 26
Baptist Mission, the, 269
Barrel rolling, 256
Barrister, native, the, 51
Basamba custom, 144, 145
Basel Mission, the, 47, 252, 269, 285
Bashilele tribe, the, 29
Basongo, 180
Batanga, 32
Batanga canoes, 13
Batetela tribe, the, 27, 178, 198
Bathurst, 294
Belgian Congo, the, 203
Belgian Congo, commerce in, 92, 154

Bell, Rev. J., 281
Benguella, 172
Benin Rubber Plantations, 242
Bennett, E. N., 96
Bihean Slave Traders, 197
Blaize, Richard, 283
Bocage, M. Du, 187
Boela tribe, the, 33
Boma, 170
Bopoto, 26
Bridewells, 176
Brine bath, the, 231
British and Foreign Bible Society the, 283
" Bula Matadi," 152
Burial customs, 25
Burroughs Wellcome and Company, 81, 83
Burtt, Joseph, 186
Bushongo tribe, the, 28

C

Cadbury—*Standard* Trial, the, 186
Calabar Missions, 280
Cameroons cocoa, 248, 251
Camwood production, 33
Canoes, dugout, 12
Cape Railway, the, 171
Capital A, The, 195
Carrier, the, 4, 151
Carson, Sir Edward, 186
Castelin, Father, 207
Catholic missionaries, Roman, 265
Chapel farms, 250
Charms, 37
Chevalier, Dr., 237, 239, 241
Chinese Labour, 135